BUILDING
BIOTECHNOLOGY

BUILDING BIOTECHNOLOGY

Business • Regulations • Patents • Law • Politics • Science

Third Edition

Yali Friedman, Ph.D.

**LOGOS
PRESS**

BUILDING BIOTECHNOLOGY
Third Edition
by Yali Friedman, Ph.D.

Published in The United States of America
by
think**Biotech LLC,** Washington, DC
WWW.BUILDINGBIOTECHNOLOGY.COM
INFO@BUILDINGBIOTECHNOLOGY.COM

10 9 8 7 6 5 4 3 2 1

ISBN-13
Hardcover: 978-0-9734676-5-9
Softcover: 978-0-9734676-6-6

Library of Congress Cataloging-in-Publication Data

Friedman, Yali.
  Building biotechnology: business, regulations, patents, law, politics, science / Yali Friedman. -- 3rd ed.
      p. cm.
  ISBN 978-0-9734676-5-9 (hardcover) -- ISBN 978-0-9734676-6-6 (softcover)
  1. Biotechnology industries. I. Title.
  HD9999.B442F75 2008
  660.6068--dc22
                            2008022042

To my family, who have inspired, motivated, and
supported me.

Contents

Introduction

Science

Laws, Regulations, and Politics

Regulation   137

Politics   171

The Business of Biotechnology

Biotechnology Company Fundamentals   183

Finance   209

Research and Development   241

Marketing   263

Licensing, Alliances, and Mergers   289

Managing Biotechnology   315

International Biotechnology 335

Conclusion

Building Biotechnology 353

Investing 367

Career Development 383

Final Words 389

Appendices

Internet Resources 393

Annotated Bibliography 399

Glossary 405

Index 423

Figures and Tables

Boxes

Preface

The benefits of genetic modification far outweigh the
hypothetical and sometimes contrived risks claimed by its
detractors.
Dr. Patrick Moore, Co-founder of Greenpeace

This book is the result of more than a decade spent investigating the biotechnology industry. My interest in biotechnology started when, as a biochemistry graduate student, I found an opportunity to create and manage one of the first websites on the business of biotechnology. Taking a biochemist's approach, I began dissecting the biotechnology industry to isolate the key elements and study their interactions. While web pages are excellent at conveying short notes and discrete topics it quickly became apparent that there was a strong need to compile my investigations in a different format—a book with comprehensive and integrated coverage of the business of biotechnology. With the desire to delve deeper into the key drivers of the biotechnology industry and provide greater coverage of the interrelation of its disparate elements, the first edition of *Building Biotechnology* was produced in 2004. Subsequent editions have seen the text grow substantially as the industry has undergone significant changes and new topics have been added.

Because the biotechnology industry is influenced by, and faces unique pressures from, scientific, legal, regulatory, political, and commercial factors, the onerous challenge of merging the respective contributions of each of these disparate domains was critical in writing this book. *Building Biotechnology* is presented in five sections: a general introduction; the science of biotechnology; legal, regulatory, and political issues; the business of biotechnology; and, a conclusion.

The scientific, legal, regulatory, and political issues are presented prior to the business fundamentals because in order to understand the business of biotechnology it is necessary to first understand how these factors shape the industry and make the business of biotechnology different from other industries. Many issues, such as drug development, are described in more than one section, providing different contexts on their fundamentals and practice.

The final section ties together the material from the previous four sections and provides additional commentary on how to engage in biotechnology business development, considerations in developing an investment strategy, and career development guidance. A comprehensive set of appendices follow, containing Internet links, an annotated bibliography, and a detailed glossary.

Several special considerations have been included to promote accessibility. Individual biotechnology companies and products are referenced in different examples and anecdotes to reinforce the concepts presented. Extensive cross-references are also included throughout the text for those readers taking a "cafeteria approach" and reading the chapters out of sequence. The annotated bibliography and detailed glossary facilitate continued learning for interested readers.

I hope that by breaking down the biotechnology industry to its key drivers and by providing numerous case studies, you will develop an appreciation of the independent and combined scientific, legal, regulatory, political, and commercial influences that define the scope of commercial biotechnology. I welcome your comments, suggestions, and questions at www.BuildingBiotechnology.com or via email at info@BuildingBiotechnology.com.

– Yali Friedman, Ph.D.

Acknowledgments

The challenge of assembling a book like *Building Biotechnology*, with broad coverage of the dynamic biotechnology industry, is immense. Despite avidly following the industry and writing and publishing on the business of biotechnology for many years, a book such as this one can only be assembled through the assistance of a diverse array of industry experts and insiders. I owe great gratitude to my colleagues and to my active readers, who contributed through their perspectives, suggestions, resources, contribution of case studies, and especially informative interactions with their constituents.

With apologies in advance to any individuals accidentally excluded, here is a list of individuals who have helped in the development of this text over its three editions:

Stephen Albainy-Jenei, Mary Ellen Clark, Peter S. Cohan, Richard Conroy, Barry Datlof, Nathaniel David, Omar Duramad, Darren Fast, Spencer Feldman, Patrik Frei, Sandy Graham, Judith Kjelstrom, William Haseltine, James Hatch, Neil Henderson, Jonathan Jacobs, Jackson Janes, Mak Joshi, Amit Kumar, Marco Landwehr, Lynn Johnson Langer, Andrew Marshall, Barry J. Marenberg, Joseph Ogden, C. Omprakash, Raymond Price, Stephen Sammut, Leo Singh, Kian Toung, Christian Walker, Fred Wilson, C. Richard Wobbe.

I

Introduction

Chapter 1

Introduction

> The ability to manipulate the genetic codes of living things
> will set off an unprecedented industrial convergence: farmers,
> doctors, drug-makers, chemical processors, computer and
> communications companies, energy companies, and many
> other commercial enterprises will be drawn into ... what
> promises to be the largest industry in the world.
> *Juan Rodriguez and Ray A. Goldberg, March 2000 Harvard*
> *Business Review*

Biotechnology inventions and products are changing paradigms in healthcare, agriculture, and industrial processes. Great opportunities exist for those who have the technologies, skills, and perseverance to bring new biotechnology products to market. These opportunities stem from the disruptive effects of biotechnology on existing markets (and the ability to create new markets), but they are tempered by a unique set of scientific, regulatory, political, economic, social, and commercial influences. Understanding the dynamic and linked contributions of the interdisciplinary array of factors affecting the commercialization of biotechnology is essential to operate in the biotechnology industry.

The biotechnology industry is not defined by a set of products or services, but by a set of enabling technologies. Whereas the literal definition of biotechnology encompasses everything from traditional agriculture to soap-making, modern definitions describe applications relying on more complex and sophisticated techniques such as genetic engineering and other forms of directed modification of living things. This book defines biotechnology as the application of molecular biology for useful purposes. This distinction is important,

3

Table 1.1 What is biotechnology?

Product / Service	Description
Biodegradable plastics	Reduce environmental impact of consumer goods
Diagnostic tests	Determine human predispositions to disease
DNA analysis	Determine paternity and assist in forensics
Genetic testing	Assist in traditional plant and animal breeding
Genetically modified crops	Improve yields and nutritional properties
Industrial enzymes	Improve efficiency and reduce environmental impact of industrial processes
Therapeutics	Treat and cure diseases

because whereas inclusion of traditional activities describes processes with established markets and mature technologies, the focus on modern techniques reflects the innovative and revolutionary possibilities of molecular biology: manipulating living organisms and parts of living organisms to capitalize on scientific discoveries, to improve upon existing solutions, or to serve new markets.

Biotechnology has applications in health, agriculture and farming, environmental remediation, and industrial processes. Within the diversity of biotechnology applications, there are two basic modes of development: products and services. Certain drugs, such as those produced in bacteria, yeast, and mammalian cells, are examples of biotechnology products (the distinction between biotechnology-derived and traditional pharmaceutical drugs is discussed in greater detail in Chapter 4). Drugs, and biotechnology research tools that are sold to pharmaceutical and other biotechnology firms, are also examples of products. Services can be sold to research firms or to companies further down value-chains for downstream application. Genetic testing is an example of a biotechnology service and is used to determine parentage, to resolve identity issues in criminal cases, and to screen for predispositions to disease.

The possible applications of biotechnology are defined by current scientific knowledge and abilities, and by the capacity of companies to develop marketable solutions from current knowledge or through

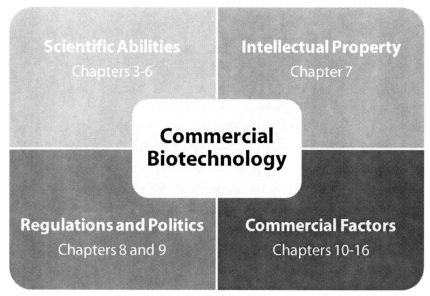

Figure 1.1 *The four pillars of biotechnology*

additional research. The commercialization of biotechnology applications is further promoted and limited by numerous legal, regulatory, and political factors. Patents serve both as a barrier to entry and an incentive for development. Changes in patent law can have profound implications on the ability of biotechnology firms to operate profitably and to obtain financing. Approval from bodies such as the Food and Drug Administration, the Department of Agriculture, and the Environmental Protection Agency is also required before many biotechnology products can be marketed or even tested. In addition to controlling the application of biotechnology, special governmental incentive programs can also motivate the development of applications that might not otherwise be commercially attractive.

Beyond these fundamental factors which define the possible applications of biotechnology, commercial factors also play an important role, as biotechnology ventures must ultimately be profitable. Whether structured as a for-profit company or a non-profit entity supported by donations or government grants, any biotechnology venture lacking an income stream cannot be sustained. Survival requires filling a need for which some party is willing to pay.

Chapter 2

The Development of Biotechnology

In science the credit goes to the man who convinces the world,
not the man to whom the idea occurs first.
Sir Francis Darwin

The modern biotechnology industry is built upon knowledge and techniques developed in the pharmaceutical industry, which employed biological extracts, dyes, and complex organic and chemical mixtures to produce drugs.

The emergence of the pharmaceutical industry is partially attributed to the development of aspirin, a drug that was developed by the German industrial chemist Felix Hoffman in 1897 and is still commonly used today. Many patients, including Hoffman's father, could not tolerate the stomach irritation associated with sodium salicylate, the standard anti-arthritis drug of the time. Armed with the knowledge that acidity associated with salicylates caused stomach discomfort, Hoffman sought a less-acidic formula and eventually produced acetylsalicylic acid, or aspirin.

As medical knowledge advanced, a focus on symptom-based treatment of diseases replaced techniques such as bloodletting and led to research on the effects of medicines and the use of defined substances as drugs. The emergence of a rational basis for medicine supported research on human biology based on the belief that a better understanding of human biology would lead to better medicine. At the same time, improved knowledge of microorganisms related to human health led to an understanding of the causes of infectious diseases and allowed new treatment paradigms. Penicillin, for example, was identified as a potential anti-infective drug based on the observation of its ability to prevent the growth of bacteria in laboratory experiments.

The growth of the pharmaceutical industry paralleled advances in knowledge of general biology and advances in methods to study and manipulate biological systems. The emergence of refined tools permitted a more fundamental study of biology—molecular biology—focusing on the fundamental processes affecting biology. The discovery of the structure of DNA in 1953 was instrumental in developing an understanding of how genetically inherited characteristics are passed from generation to generation.

The first biotechnology companies were formed in the 1970s and 1980s. Knowledge of the molecular fundamentals of biology and development of tools to manipulate biological systems laid the foundation for the biotechnology industry, which employs the directed application of molecular biology for useful purposes. Biotechnology drug development not only uses methods and strategies different from traditional pharmaceutical development, it also produces different products. By selecting proteins such as insulin and erythropoietin, whose functions were already known, as their lead compounds, firms such as Amgen, Genentech, Chiron, and Genzyme employed a directed drug design strategy. In contrast with the chemical synthesis and biological extraction techniques that produced traditional pharmaceutical drugs, these early biotechnology companies used recombinant DNA techniques that enabled them to produce proteins as therapies (see Chapter 4 for more details).

KNOWLEDGE AND SKILLS

A brief history of selected Nobel Prize awards in the categories of Chemistry, and Physiology or Medicine provides a path to follow the scientific developments that spawned the biotechnology industry. Nobel Prizes are awarded for outstanding achievements and contributions and are internationally recognized as the most prestigious awards in the fields for which they are awarded. Because it can take some time for the significance of a discovery to emerge, many Nobel Prizes are awarded years after the actual discovery.

Frederick Sanger was awarded the Nobel Prize in Chemistry in

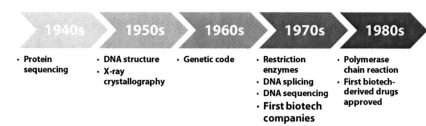

Figure 2.1 *Knowledge and skills enabling biotechnology*

1958 for his determination of the protein sequence of insulin. Sanger, who began his mission in 1943, developed numerous techniques to directly sequence proteins, which enabled scientists to better understand these biological molecules. Knowledge of the sequence of human insulin enabled Genentech to develop recombinant human insulin—the first biotechnology drug—in 1982.

Between 1950 and 1956, Herbert Hauptman and Jerome Karle laid the foundations for the development of X-ray methods to determine the structure of crystallized molecules. They shared the 1985 Nobel Prize in Chemistry for their work. X-ray crystallography determines a molecule's three-dimensional structure by analyzing the X-ray diffraction patterns of crystals of the molecule. The complexity of organic molecules such as DNA and proteins meant that many structures were not known until the advent of X-ray crystallography. X-ray crystallography aided discovery of the structure of DNA in 1953, a significant advance in molecular biology that set the stage for modern biotechnology. James Watson, Francis Crick, and Maurice Wilkins shared the 1962 Nobel Prize in Physiology or Medicine for their work in discovering the structure of DNA. This discovery enabled elucidation of the mechanisms for control of gene expression and hereditary transfer of genetic information.

Following the discovery of the structure of DNA, the need to explain its role in cellular functions remained. Robert Holley, Har Gobind Khorana, and Marshall Nirenberg shared the 1968 Nobel Prize in Physiology or Medicine for their contributions in deciphering the genetic code, the language by which information is contained in DNA, and for elucidating how this information is translated by cells.

Werner Arber, Dan Nathans, and Hamilton Smith shared the

Nobel Prize in Physiology or Medicine in 1978 for the discovery of restriction enzymes and their application to problems of molecular genetics. It was the pioneering work of these three scientists that enabled development of the DNA manipulation techniques that permitted Stanley Cohen and Herbert Boyer to develop methods for splicing DNA from different sources, often referred to as recombinant DNA (rDNA) technology.

The 1980 Nobel Prize in Chemistry was awarded to Paul Berg, Walter Gilbert, and Frederick Sanger. Berg was recognized for his "fundamental studies of the biochemistry of nucleic acids, with particular regard to recombinant-DNA," and Gilbert and Sanger for their "contributions concerning the determination of base sequences in nucleic acids." The ability to determine the sequence of DNA was central to the Human Genome Project and is a key element in biotechnology research and development.

Kary Mullis and Michael Smith shared the 1993 Nobel Prize in Chemistry for their respective development of the polymerase chain reaction (PCR), and site-directed mutagenesis. Mullis' PCR permits the specific production of copies of a specific DNA segment, even in the presence of a complex mixture of DNA. This technique has applications in forensics, paternity and heritage testing, medical diagnostics, archaeology and anthropology. Application of PCR and site-directed mutagenesis permits the directed modification of genetic sequences, effectively reprogramming genes.

APPLICATION

The significant scientific developments described above set the stage for the biotechnology industry. Understanding the role of DNA in programming the abilities of individual cells, combined with knowledge of how information is encoded in DNA, the mechanisms by which cells use this information, and the development of molecular biology techniques to manipulate DNA, gave rise to modern biotechnology.

In 1973, Stanley Cohen at Stanford University and Herbert Boyer at the University of California at San Francisco developed methods to splice genes and express foreign proteins in bacteria. This made it possible to deliberately make defined changes to biological systems, permitting the directed modification of microbes and cell cultures to produce desired products. Boyer and venture capitalist Robert Swanson formed Genentech in 1976, a defining event in modern biotechnology. Genentech, one of the first biotechnology companies, aimed to commercialize gene splicing technology by initially producing recombinant human insulin in bacteria to treat diabetes.

Prior to 1976, drugs were either chemically synthesized or extracted from living sources. Before bacterial production, insulin was commonly extracted from pig pancreas and required the sacrifice of 50 animals to produce sufficient insulin for a single person for one year. The advent of gene splicing introduced new possibilities, facilitating drug development without screening libraries of chemicals and biological extracts, and enabled scientists to select proteins whose function was already known as lead compounds.

Following proof-of-principle production of a neurotransmitter, Genentech produced recombinant human insulin in bacteria in 1978, later to become the first recombinant DNA drug approved by the Food and Drug Administration.

In 1980, prior to FDA approval of its recombinant human insulin, Genentech capitalized on positive market sentiment towards biotechnology and raised $35 million in an initial public stock offering. Without the resources to fully develop and commercialize recombinant human insulin as a drug, Genentech had licensed manufacturing and distribution rights to Eli Lilly, the dominant supplier of beef and pig insulin. Aiming to independently develop and commercialize a drug, Genentech became the first biotechnology company to market its own biopharmaceutical product in 1985 when it used gene splicing to produce human growth hormone, a drug previously available only by harvesting pituitary glands from deceased human organ donors. Since then, Genentech has produced many additional products, was bought by Roche Pharmaceuticals, and was subsequently resold on the public markets.

Genentech focused on one of the first core technologies defining the biotechnology industry, but is not the first biotechnology company. That status belongs to Cetus. Cetus was founded in Berkeley, CA, in 1971 and initially focused on using automated methods to screen for microorganisms with industrial applications. Despite developing the Nobel Prize-winning polymerase chain reaction technology, the company was not able to maintain independence, and was acquired by Chiron in 1991 (see Box *Cetus spreads itself too thin* in Chapter 12).

COMMERCIALIZATION

The history of Genentech serves as a paradigm for biotechnology product development and corporate growth. Genentech was founded to exploit a novel scientific innovation. Without sufficient resources to fully develop and commercialize its first product, Genentech licensed these rights to a larger partner. Tapping revenues from early products enabled Genentech to develop sufficient bulk to fully research, develop, and commercialize its own products.

The means and motivation must exist in order to develop a biotechnology product. The motivating factor can be as simple as consumer demand, permitting a company to derive revenues from sales. Alternatively, if a technology is sufficiently appealing, the potential to create new markets can motivate development. Conversely, public resistance to biotechnology products, such as opposition to genetically modified crops, can exert a negative influence on the marketability of a product. Whether a company is compensated directly from sales, government grants, or awards, or is compensated indirectly from tax credits, there must be some motivation to support development.

Legal and regulatory pressures can promote or discourage development. Long development times and the relative ease of reverse-engineering necessitate intellectual property protection for biotechnology products. Patents grant the right to exclude others from practicing an invention, providing an incentive for patent holders or licensees to develop patented applications by preventing competitors from capitalizing on their research and development investments. For this reason, many biotechnology firms form around patented scientific methods or proprietary knowledge that create a barrier to competitors and a source of revenue through licensing of partially- or fully-developed products and technologies.

Box
Genentech: Commercializing a new technology

Genentech was founded in 1976 to capitalize on the revolutionary gene splicing technology developed by Stanley Cohen and Herbert Boyer. The company has since diversified to other technologies and boasts revenues in excess of $11 billion. It is also the only biotechnology company to never trade below its initial public offering (IPO) price, and has been profitable for all but two of its years as a public corporation.

1976: Genentech founded by Robert Swanson and Herbert Boyer

1977: Genentech produces the first human protein (somatostatin) in a microorganism

1978: Genentech produces human insulin in bacteria

1979: Genentech produces human growth hormone in bacteria

1980: Genentech goes public and sets a record for IPO stock price appreciation

1982: Genentech launches the first recombinant DNA drug: human insulin (licensed to Eli Lilly)

1985: Protropin (human growth hormone) approved by FDA—the first recombinant pharmaceutical product to be manufactured and marketed by a biotechnology company

1986: Roferon (interferon alpha-2a) approved by FDA and licensed to Hoffmann-La Roche

1987: Activase (tissue-plasminogen activator) approved

1990: Genentech's Hepatitis B vaccine, licensed to SmithKline Beecham, receives FDA approval

1993: Nutropin (somatropin) receives FDA approval
Pulmozyme receives FDA approval

1997: Rituxan receives FDA approval

1998: Herceptin (trastuzumab) receives FDA approval

1999: Roche exercises option to purchase ownership of Genentech, offers 22 million shares in an IPO six weeks later in the largest healthcare-related IPO ever
Secondary offering releases another 20 million shares, raising $2.87 billion in the largest secondary offering ever

2000: Third offering of 19 million shares raises $3.1 billion

Box

Amgen: Capitalizing on innovation

Amgen was founded to capitalize on expanding opportunities in biotechnology. Founding CEO George Rathmann willingly left legacy pharmaceutical company Abbott for the more open and free environment of a biotechnology start-up. Today Amgen leads the biotechnology industry with revenues in excess of $14 billion and more than 20,000 employees.

1980: Amgen formed as Applied Molecular Genetics by a group of scientists and venture capitalists with a $19 million private-equity placement from venture capital firms and two major corporations
Employees: 3

1981: Amgen begins operations in Thousand Oaks, CA
Employees: 31

1983: Amgen isolates gene for human erythropoietin (EPO), later to become Epogen
Employees: 124

1985: Amgen sells Johnson & Johnson partial rights to EPO while still in development
Research team isolates gene for granulocyte colony-stimulating factor (G-CSF), later to become Neupogen
Employees: 196

1987: Amgen receives first patent for Epogen
Employees: 344

1989: Amgen receives first patent for Neupogen
FDA approves Epogen as orphan drug
Employees: 667

1991: FDA approves Neupogen
Employees: 1,723

1992: Amgen sales surpass $1 billion
Employees: 2,335

1996: Amgen sales surpass $2 billion
Employees: 4,646

1999: Amgen sales surpass $3 billion
Employees: 6,342

2002: Amgen acquires Immunex (see Box *Enbrel: Underestimating market demand* in Chapter 13)

A characteristic distinguishing biotechnology (and pharmaceutical) products from those of many other industries is the requirement for rigorous and lengthy assessments to verify the safety and, in the case of drugs, efficacy, of products prior to being able to market them. Companies and financiers are therefore often unwilling to commit resources for development of drugs and other products for which the regulatory path is uncertain.

In addition to limiting development, government regulations can also motivate development. The Orphan Drug Act is an example of an incentive for drug development; tax credits and market exclusivity are granted to companies developing drugs for small populations that meet specific criteria.

Biotechnology development is fueled by innovation. The importance of specialized knowledge means that entrepreneurship by accomplished scientists is common in the genesis of biotechnology companies. The significant risk of product development failure compels biotechnology companies to focus on research and development until marketable products emerge. Patents and other barriers to entry are essential to prevent late-entering competitors from capitalizing on the efforts of pioneers.

INDUSTRY TRENDS

Many of the companies founded in the 1970s and 1980s sought to become fully vertically integrated drug developers, incorporating processes from drug discovery and development through production and sales. The prototypical company of this era aimed to develop treatments for unmet disease conditions and used the financing power of favorable public markets to fund expensive drug development efforts. Companies such as Genentech and Amgen were successful enough to achieve independence, but when market support for biotechnology disappeared, many companies had to reformulate their business models, merge, or liquidate.

Two impediments that prevented many of these early biotechnology companies from achieving vertical integration were the limited amount of available funding, which could not support the number of high-burn companies being founded, and the lack of experienced managers. The number of biotechnology companies aiming to become fully integrated diluted the amount of funding available at the time, limiting the support that each company could attain. Additionally, in order to develop vertically integrated companies, young startups needed managers with broad expertise from product development to commercialization. The only potential source for people with these skills was the pharmaceutical industry. Unfortunately, the pharmaceutical industry had divided the drug discovery and commercialization process into separate divisions managed by specialists, so no suitable managers existed. Furthermore, because biotechnology companies were seen as competitors, established pharmaceutical companies had little incentive for collaboration. By the late 1980s, pharmaceutical company sentiment towards biotechnology partnerships softened as pharmaceutical companies found themselves unable to maintain their growth rates solely by their internal research programs.

The 1990s saw the emergence of platform and tool-based companies seeking to commercialize drug targets, services, and technologies that could be sold or licensed to other companies. Revenue streams emerged from partner licensing fees, royalties, and research contracts. Although revenues from tools and services can make a company profitable, there is always the risk that these offerings can become commodities or obsolete.

Recognizing that revenues from tools and services could fund product development efforts, hybrid business models emerged in the late 1990s and early 2000s, capitalizing on the stability of tool and service sales while still selling the promise of product development. In addition to licensing or selling research tools to others, they were also used internally for product development. In principle, hybrid companies could therefore enjoy stable revenues from licensing and sales agreements while attracting investors by selling the promise of product development. The time and energy that must be devoted to marketing and selling tool offerings and keeping them current can make product development slower for hybrids than for product-focused companies. This reduced pace is balanced by the stability granted by revenues derived from tools which permit hybrid companies to better weather unfavorable financing environments.

The "no research, development only" (NRDO) model gained favor in the wake of the biotechnology bubble of 2000. A derivation of the specialty pharmaceutical model of seeking additional markets for drugs already approved in one or more countries, the goal of NRDO firms is to acquire promising lead compounds and manage their clinical trials, at which point the drugs can be marketed in partnership with, or sold to, larger firms. NRDO firms were able to capitalize on the wealth of drug leads and managers that could be inexpensively acquired from firms struggling or liquidating as a result of unfavorable market conditions. A limitation of the NRDO model derives from the reality that many important discoveries in science emerge in the course of unrelated research. By not participating directly in research, NRDO firms are unable to realize the significant upside of tangential discoveries that emerge from research. A lack of internal drug development talent also challenges managers to obtain skilled guidance, often from paid consultants or contract research laboratories rather than internal experts, to assess the quality of potential product acquisitions.

Another recent trend is the move toward larger-scale projects. The ability to automate procedures such as DNA sequencing, microarray analysis, and drug screening make it possible to perform research at an unprecedented scale. Data mining and massive bioinformatics projects have also formed the core of companies. This shift in scale demonstrates a very important change in the way research is conducted. The ability to perform large-scale experiments requires reliability and automation, attributes not often found in basic scientific discoveries and methods. DNA sequencing, a procedure that can now be fully automated, once required days of manual labor. Just as computers have advanced knowledge in other disciplines with their ability to process information and reliably and repeatedly perform tasks, the ability to automate biotechnology experiments will lead to greater discoveries at lower costs.

II

Science

Scientific research is a slow, painstaking process often fraught with setbacks; unfortunately, managers unfamiliar with this process fail to appreciate these difficulties. Because biotechnology involves novel products and techniques, it is difficult to predict the hurdles that will be encountered or the precise outcome of development efforts. Furthermore, a regulatory burden arises from the need to verify the safety and efficacy of biotechnology products. This section presents a detailed overview of relevant scientific topics to facilitate better understanding of challenges and opportunities of biotechnology research.

The biotechnology industry is not defined by a set of products, but by a set of enabling technologies. The prototypical biotechnology company focuses on research and development and uses molecular biology techniques to develop drugs and other useful products. Molecular biology is distinguished from general biology by the fundamental nature of the material studied. Whereas biology is the general study of life, molecular biology seeks to understand the inner workings of life's processes. Using molecular biology techniques, biotechnology companies are able to manipulate the fundamental processes responsible for diseases, or tap biology for other useful purposes.

Biotechnology companies engage in basic and applied research (see Figure 4.3). Basic research is primarily focused on acquiring new knowledge regarding the principles underlying phenomena and observations. Basic research is characterized by hypothesis testing, analytical experiments, and theory development. Building on basic research, applied research develops new knowledge and applications. Biotechnology firms use applied research to develop and commercialize the innovations and discoveries that emerge in the course of basic research.

Prior to the advent of molecular biology, biologists sought to answer such questions as how our physical characteristics are inherited from one generation to the next, how food is converted into energy, and how different cell types develop and perform their specialized roles. These researchers were able to identify agents responsible for disease and the role of human tissues in health and disease. It was not until the development of molecular biology that it became possible to discover and alter the actual processes responsible for health and disease states.

In 1953, Francis Crick and James Watson discovered the structure of DNA, the primary source of information in cells that permits genetic characteristics to be passed on from one generation to the next and bestows traits on cells. Following elucidation of the structure and function of DNA, the genetic code by which information is stored in genes was deciphered and the methods by which this information is ultimately translated were determined. These developments helped redefine biological research, but it took nearly 30 years—the first biotechnology drug was approved in 1982—for modern biotechnology to demonstrate its potential.

Understanding the fundamentals of molecular biology, combined with the ability to introduce genes into organisms, enabled biotechnology: the directed modification of living things toward useful ends.

As the science has matured, companies in previously unrelated industries have invested increasingly in biotechnology research. The application of biotechnology in diverse industries makes it difficult to define biotechnology companies discretely; every company that uses biotechnology is not a biotechnology company. Instead of being defined solely by their research activities, biotechnology companies are defined by the concentration of their focus on biotechnology research and development.

The application of biotechnology provides new answers to old problems, but also introduces new challenges. Markets for many applications are well established. The question in these cases is not whether the customers exist, but if it is possible to produce a useful and compelling product at a reasonable cost in a reasonable amount of time.

Chapter 3

Introduction to Molecular Biology

Everything should be as simple as possible, but not simpler.
Albert Einstein

B iotechnology research seeks to develop applications of molecular biology. Many sources use analogies to recipe books or blueprints to explain the role of DNA and genes in molecular biology. Ultimately, these analogies obscure the importance of topics such as regulation of gene expression, which is of fundamental importance in understanding molecular biology. When applying one's knowledge of biotechnology fundamentals, most metaphors fail. It is only by understanding molecular biology and biotechnology applications that one can appreciate the applications and limitations of techniques used in molecular biology.

This chapter presents a brief, analogy-free, introduction to molecular biology. Subsequent chapters describe the tools, techniques, and applications of biotechnology and provide greater details on the potential and limitations of molecular biology.

INFORMATION FLOW IN MOLECULAR BIOLOGY

In order to understand the basis of most biotechnology applications, it is necessary to first understand the process by which information in genes leads to the formation of structural and functional proteins.

Proteins serve structural and functional roles that give individual cells—and by extension whole organisms—specific structures and functional characteristics. When many people think of proteins,

Information Flow in Molecular Biology

Genetic information is contained in DNA and leads to the formation of proteins through an intermediary called mRNA.

they think of foods such as meat and beans. While animal muscle and plant seeds are excellent sources of dietary protein, proteins play a central role in all cell types and perform functional and structural roles (see Table 3.1). Examples of structural proteins include keratin, which makes skin waterproof, and myosin, which interacts with other proteins in muscles to make them flex.

DNA contains information that describes the construction of proteins. The process of protein synthesis is as follows:

1. DNA contains the information to produce proteins.
2. Information encoded in DNA is *transcribed* into a molecule called messenger RNA (mRNA)—effectively a "working copy" of the DNA sequence of a given gene.
3. mRNA is *translated* into proteins by the protein synthesis machinery, the composition of the resulting protein corresponding to the original DNA instructions.

This basic mechanism is conserved in all life forms, from bacteria to humans. The implication of this common process that converts information in DNA into functional proteins is that similar techniques can be used to investigate and manipulate all biological systems. Furthermore, it is possible to make human therapeutic pro-

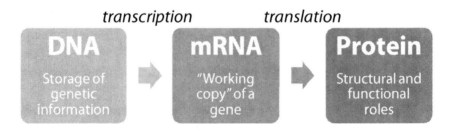

Figure 3.1 *Simplified model of information flow in molecular biology*

teins, for example, in organisms as distantly related as bacteria.

Understanding the roles of DNA, RNA, and protein and their relationships to each other is essential to understanding molecular biology. While there are some specific exceptions (e.g., retroviruses and prions) to the order and direction of information flow shown in Figure 3.1, these examples still fit within the general framework, and the majority of biological systems use the framework as presented.

DNA: STORING AND RELAYING INFORMATION

Deoxyribonucleic acid (DNA) is the primary source of genetic information in cells. Humans, plants, animals, and bacteria all contain DNA. DNA is physically passed from generation to generation, bestowing certain traits of parents to their children. The reason why children have physical characteristics from each of their parents—a child may have their mother's eye color and father's hair color—is

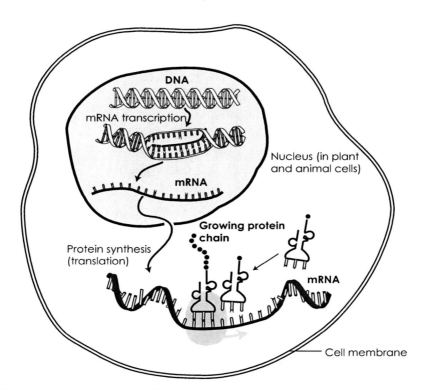

Figure 3.2 *General scheme of gene expression*
Modified from National Human Genome Research Institute

because they received half their DNA from each parent.

Each of our cells (with a few exceptions like red blood cells, eggs, and sperm) contain all the DNA required to code our genetic features. Individual regions of DNA that confer traits are called genes. Information in genes is relayed to the protein synthesis machinery within cells where it dictates the production of proteins. The word "genome" refers to all the DNA in an organism. The human genome contains approximately 30,000 genes arrayed on 46 long stretches of DNA called chromosomes.

DNA is essentially composed of two intertwined strands that form a double helix. The two strands of DNA are said to be complementary because the sequence of one strand indicates the sequence of the opposite strand, like a photograph and its negative. Each strand is physically composed of four different chemical units called nucleotides, the sequence of which encodes the genetic information. These four chemical units, adenine, cytosine, guanine, and thymine, are often abbreviated as A, C, G, and T, respectively. Just as the English language can be expressed in twenty-six letters, the genetic code is expressed in these four chemical units. A DNA "sequence" refers to the specific order of A's, C's, G's, and T's in a stretch of DNA.

There are two essential elements of genes: coding and regulatory elements. The coding elements of genes are first transcribed as mRNA, which is then translated into protein. The chemical sequence of A's, C's, G's, and T's in the coding region of a gene determines the composition and structure of the resulting protein and, by extension, its function. Regulatory elements affect the rate at which genes are transcribed and translated, and may be interspersed within the coding sequence or outside of it. Regulatory elements also control the cell types within which specific genes are activated, and the timing and magnitude of gene expression. Gene regulation thereby allows individual proteins to be expressed only in certain cells at specific times and at specific rates.

Proper regulation of gene expression—the production of gene products—is essential. Under- or over-expression of genes can have deleterious effects. For example, many forms of cancer are caused by mis-regulation of gene expression that results in uncontrolled cell division. A potential solution for diseases resulting from low expres-

Nucleus (not present in bacteria)

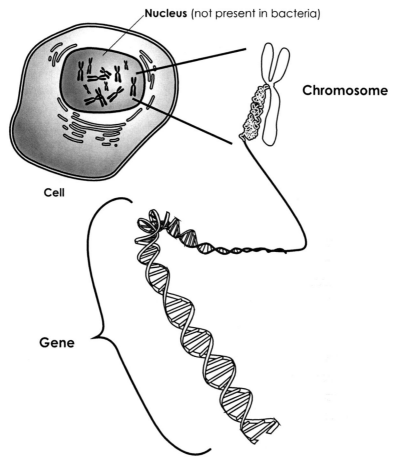

Chromosome

Cell

Gene

Figure 3.3 *DNA: Chromosomes and genes*
Modified from National Human Genome Research Institute

sion of genes is to use gene therapy to introduce affected genes or regulatory elements to spur additional production. One of the challenges of gene therapy is developing methods to regulate the expression of genes that are introduced into cells and ensure that they are not over-expressed. A solution for diseases caused by over-expressed genes is RNA interference. This procedure prevents translation of mRNA, inhibiting protein production. RNA interference is described in further detail in Chapter 6.

Box
Human chromosomes and genetic trait inheritance

The human genome is composed of chromosomes. We get 23 chromosomes from our mother and 23 chromosomes from our father, constituting 23 pairs. While 22 of the 23 chromosome pairs are similar in both men and women, the 23rd pair is quite different and determines the sex of an individual. For the 23rd pair of chromosomes, women have two X-chromosomes while men have one X- and one Y-chromosome. Because X-chromosomes contain more DNA than Y-chromosomes, they are physically larger than Y-chromosomes. Having too many or too few chromosomes can affect gene regulation and cause diseases. Down's Syndrome, for example, occurs in individuals with three copies of chromosome 21.

The roles of X- and Y-chromosomes are important in understanding sex-linked diseases. Women do not have Y-chromosomes, so diseases that are caused by defective genes on the Y-chromosome can only occur in men. Additionally, men only have one X-chromosome, so mutations in genes on the X-chromosome are more likely to affect males, because the second X-chromosome in women can sometimes compensate for mutations on the first. Color blindness, caused by a mutation on the X-chromosome, is more common in men than women for this reason.

mRNA: THE MESSENGER

Messenger RNA (mRNA) is used to relay information from genes in DNA to the protein synthesis machinery. An additional feature of mRNA is that it can be destroyed once sufficient protein is produced, permitting an extra level of control of gene expression. RNA is also present in forms other than mRNA, some of which are described later in this chapter.

It is possible to affect expression of genes by targeting their mRNA with antisense RNA or DNA—nucleic acids which can bind the mRNA. Unlike DNA, which is usually double-stranded, mRNA is single stranded. Nucleic acids (DNA or RNA) containing a sequence that can bind to a given mRNA will prevent translation by the protein synthesis machinery, inhibiting gene expression. The Flavr Savr tomato, a tomato engineered to have a long shelf life, was produced by introducing antisense RNA corresponding to mRNA for an enzyme involved in fruit spoilage. Inhibiting expression of this gene delays spoilage. In 1998 the FDA approved Isis Pharmaceuticals' Vitravene, the first antisense drug, to treat cytomegalovirus-induced retinitis.

TRANSLATION: MAKING PROTEINS

Just as DNA and RNA are composed of linked nucleotides, proteins are comprised of chains of amino acid units. When mRNA is translated to produce a protein, the protein-synthesis machinery "reads" the nucleotides three at a time, assembling amino acid chains that correspond to the mRNA sequence. The basic elements of the protein-synthesis machinery are tRNA, a form of RNA that *transfers* amino acids to the protein-synthesis machinery in a way that enables them to be linked together, and ribosomes, which help form the chemical bonds that attach amino acids in a protein chain.

The three-nucleotide sequence elements on mRNA that code for individual amino acids are called codons. These are matched by

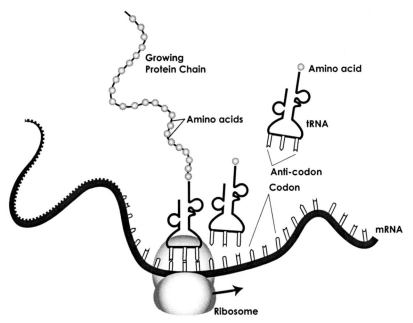

Figure 3.4 *Protein translation*
Modified from National Human Genome Research Institute

anti-codons on tRNA to ensure that the appropriate amino acid is aligned with a given mRNA sequence. The 64 possible combinations of A, C, G, and T at each codon code for only 20 different amino acids. This redundancy in the genetic code, permitting multiple codons to specify common amino acids, is considered a form of protection against DNA mutations and has applications in identifying foreign DNA from sources such as viruses which may use different "dialects" of the genetic code.

The chemical characteristics of amino acids in a protein cause it to fold into a defined 3-dimensional structure. That determines the protein's function. Because the DNA sequence of a gene dictates the sequence of amino acids in a protein, and the sequence of these acids in a protein determines its structure, one can deduce a protein sequence, and potentially its structure and function, from the gene sequence encoding it.

PROTEINS AND ENZYMES

Proteins, the workhorses of cells, are responsible for the majority of structural features and functional characteristics in cells. Enzymes are proteins that perform functional roles as part of the cellular process. Different types of cells get their characteristics by expressing a specific array of genes, resulting in production of a complement of proteins that give each cell type its unique characteristics. Pancreatic cells, for example, produce the protein insulin to regulate blood sugar levels; neurons produce neurotransmitters essential for brain function; and hemoglobin is made in blood cells, enabling them to carry oxygen. Examples of enzymes include proteases that break down proteins or enable digestion of food, and polymerases that assemble DNA and RNA. Some genes are expressed only in certain cell types whereas others are widely expressed. Examples of widely-expressed genes include those encoding proteins and enzymes involved in general cellular activities such as DNA replication, mRNA translation, protein synthesis, energy production and maintenance of structural integrity.

Production of inappropriate proteins in cell types and mis-regulation of protein expression are at the root of many diseases. As

Table 3.1 *Examples of protein and enzyme functions*

Enzyme	Function
Amylase	Breaks down starches and other complex carbohydrates into basic sugars
Cellulase	Breaks down cellulose, found in the cell walls of plants
Lipase	Breaks down fats
Protease	Breaks down proteins

Protein	Function
Collagen	Main protein in connective tissue; structural roles in skin, cartilage, teeth, bone, and other tissues
Keratin	Makes skin waterproof and contributes to strength and flexibility
Myosin	Muscle contraction

mentioned above, many cancers result from mis-regulation of gene expression that causes uncontrolled cell division.

Molecular biologists can transfer genes from humans and other animals into bacteria, yeast, and other organisms to confer the ability to produce specific proteins that may be extracted for therapeutic use. For example, Genentech produced its first drug by introducing the gene for human insulin into bacteria and extracted the resulting protein to produce a treatment for human diabetes. Genes can also be transferred from one organism to another to confer new attributes. Pesticide-resistant crops have been produced by incorporating naturally-occuring pesticidal proteins into plants. Bacteria have also been modified to perform roles such as decomposing oil spills by adding genes encoding proteins with the ability to break down components of oil. Additional examples are described in Chapter 6.

Other forms of RNA

Traditional molecular biology held that the primary role of RNA in cells was largely limited to housekeeping functions such as transferring information from DNA to the protein synthesis machinery (mRNA), transporting amino acids to be assembled into proteins (tRNA), and translating mRNA into protein (rRNA).

Sidney Altman and Thomas Cech shared the 1989 Nobel Prize

Table 3.2 Selected *RNA types*

RNA type	Function
mRNA	Messenger RNA. Contains a working copy of a gene sequence and is read by the protein synthesis machinery to produce proteins.
tRNA	Transfer RNA. Transfers amino acids to the protein synthesis machinery to produce proteins.
rRNA	Ribosomal RNA. Part of the protein synthesis machinery. Also useful for determining evolutionary similarity between organisms.
aRNA	Antisense RNA. Used for gene regulation.
siRNA	Small Interfering RNA. Used for gene regulation.
snRNA	Small Nuclear RNA. Used to edit mRNA, regulate gene expression, and maintain chromosome tips (telomeres).

in Chemistry for their discovery of catalytic properties of RNA. The ability to catalyze (increase the rate of) biochemical reactions had previously been thought to only exist in proteins. Altman and Cech found a role for RNA in the splicing of mRNAs, ultimately making it possible for a single gene to give rise to several different proteins. The significance of Altman and Cech's discovery was expanded more than a decade after they received the Nobel Prize. Following sequencing of the human genome it was discovered that the human genome contained only a fraction of the genes previously thought necessary to produce the complete set of proteins comprising human biology. The ability of this small set of genes to produce the full complement of human proteins could largely be explained through mRNA splicing.

More recently myriad forms of RNA have been discovered, and diverse roles for RNA have also been elucidated (see Table 3.2). These discoveries indicate that controlling cellular activities is more complex than previously thought, suggesting that there are also more opportunities to influence cellular activities.

THE BIG PICTURE

Genes interact with the environment and with each other to confer traits. While the presence or absence of a gene can potentially confer a given trait, environmental factors also play a role. Our physical characteristics are a combination of genetic and environmental factors. A child with a hypothetical *tallness* gene, for instance, would not necessarily grow taller than a child without the gene; the child with the *tallness* gene would also require adequate nutrition to fuel the extra growth (and the effect of the *tallness* gene may be limited or enhanced by the action of other genes). Rather than thinking of genes as determinants of physical characteristics, they should be regarded as potentials or predispositions for characteristics.

The ability to modify characteristics of cells is similarly limited by biological and physical constraints. Since some cells are rapidly replaced, induced changes will be quickly lost. Other cells are dormant, precluding their potential to express modifications.

Furthermore, biology is complicated. In fields such as industrial chemistry or engineering, applications are developed from well-characterized principles. With biotechnology on the leading edge of molecular biology research, it can be difficult or impossible to foretell the outcomes of manipulations and they can have unforeseen consequences. Because it is not possible to fully predict the outcome of these procedures, scientists must perform experiments, take observations, refine theories, and finally develop functional applications. This is why biotechnology research is so complex, time consuming, and fraught with unforeseen setbacks and disappointments.

Chapter 4

Drug Development

It is only by the means of the sciences of life that the quality of life can be radically changed.
Aldous Huxley

D rugs are substances that affect the functions of living things and are administered to treat, prevent, or cure unwanted diseases and symptoms. The United States Food and Drug Administration (FDA) regulates drug marketing, requiring manufacturers to prove their products to be safe, effective, and appropriately labeled. The drug development process identifies drug candidates and subjects them to increasingly stringent tests to assess their safety and efficacy. Drug development is paradigmatic of the general process by which biotechnology products are developed, with one important difference: non-therapeutic products are not subject to the same regulatory pressures to gain marketing approval.

Scientists start with simple, defined, model systems that enable them to identify potential drugs. These potential drugs are then tested in increasingly complex and real-world situations to prove their efficacy. It is important to test for as many contingencies as possible. Something that works well in a simple model system may fail in real-world use due to any number of unforeseen circumstances.

The description of drug development in this section is presented as a model for biotechnology product development. Producing and selling drugs consists of three basic stages: discovery, development, and commercialization. Less than 1 percent of early candidate compounds make it through the drug development process.

Discovery-stage research produces lead compounds that must

pass tests to predict their toxicity, to determine if they can be effectively administered, and to project the likelihood of recouping development costs (see Figure 4.4).

The development process builds on observations and products from discovery-stage research. Formulations are developed to optimize drug administration, and pre-clinical and clinical trials are employed to test the safety and efficacy of drugs in humans. In addition to testing the physical properties of a drug, manufacturing processes which can consistently produce doses of equivalent purity and efficacy over a period of time must be developed and tested.

The products of development-stage activities—approved drugs— are commercialized. Drug marketing, one of the dominant elements of commercialization, is described in Chapter 13.

BIOTECHNOLOGY VS. PHARMACEUTICAL DRUG DEVELOPMENT

Traditional pharmaceutical drugs differ from biotechnology-derived drugs in the methods by which they are discovered and manufactured. As a result, the resulting drugs have markedly different characteristics.

To distinguish traditional pharmaceutical from biotechnology drug development, consider the traditional pharmaceutical and biotechnology forms of therapeutic insulin. Prior to Genentech's production of recombinant human insulin, pharmaceutical companies extracted insulin from the pancreas of pigs, cows, and horses. Glands from fifty pigs were needed to produce sufficient insulin to treat a single person for one year. Insulin from these sources was subject to disease transmission, shortages, and reactions with the human immune system. Genentech produced recombinant human insulin, the first biotechnology drug, by synthesizing it in bacteria. Bacterial fermentation allowed for greater production capacity, avoidance of immune system reactions typical of non-human forms of insulin, and elimination of the threat of transmission of animal diseases. This example illustrates how the fundamentals-based approach to product development employed by biotechnology firms permits the development of solutions not attainable by traditional pharmaceuti-

Small-Molecule Drug

Biologic Drug

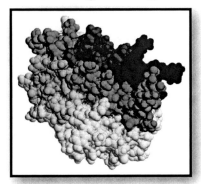

Aspirin
23 atoms

Erythropoietin
1297 atoms

Figure 4.1 *Small-molecule and biologic drugs*

cal development.

Traditional pharmaceutical drug discovery was based on trial-and-error screening of synthetic compounds and directed selection of biological extracts that can affect model systems. The emphasis of research was to understand biological systems in order to find potential drug targets. Compounds and extracts that interacted with these targets were then selected for further study to see if they could be used as drugs.

The molecular biology techniques used by biotechnology firms differ from traditional pharmaceutical development because they permit a finer-scale analysis of biological systems and the directed design of biological compounds as drug candidates. Traditional pharmaceutical development was limited to chemical synthesis and biological extracts. The reason for this limitation was that traditional pharmaceutical development originated before the advent of molecular biology techniques, which enable the directed design and production of biological molecules.

Drugs produced by traditional pharmaceutical means tend to be small molecules that are orally doseable as tablets, capsules, or liquids (see Figure 4.1). Following absorption in the gastrointestinal tract these drugs travel throughout the body in the bloodstream, and can often be mass-produced for a relatively low cost.

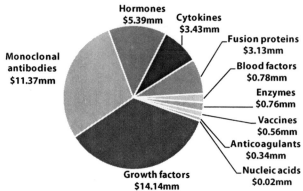

2006 Biotechnology drug sales by category

Figure 4.2 *Biotechnology drug categories*

While biotechnology research techniques do enable new possibilities, they have not rendered pharmaceutical research techniques obsolete. Traditional pharmaceutical research is still practised because of available expertise, the abundance of chemical and biological-extract libraries, and the strength of techniques for target selection and optimization.

The majority of biotechnology drugs have been proteins, such as growth factors, monoclonal antibodies, hormones, and cytokines (see Figure 4.2). Other categories include nucleic acids and vaccines. The ability to design, modify, and synthesize biological compounds means that many biotechnology drugs are larger and more complex than traditional pharmaceutical drugs.

Drug delivery is an issue for biotechnology-derived drugs because proteins and other biotechnology drugs such as nucleic acids are less likely to survive the acidic conditions in the stomach and are generally unable to pass through the intestinal lining and travel the bloodstream to their therapeutic target. Biotechnology drug delivery techniques include injection, skin patches, and inhalation. Some of these alternative delivery systems allow for more precise tissue targeting and improved dosage control, presenting new opportunities.

Because the post-discovery activities for biotechnology-derived and traditional pharmaceutical drugs are relatively similar, modern pharmaceutical firms are also able to develop lead compounds gen-

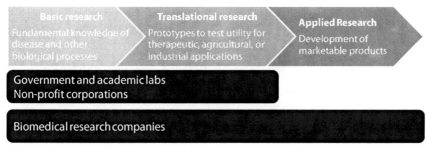

Figure 4.3 *Basic and applied research*

erated by biotechnology firms, and actively engage in biotechnology research themselves. Conversely, as the biotechnology industry has matured, some biotechnology companies have developed late-stage development and marketing abilities on par with small and medium-sized pharmaceutical companies. These shifts in the roles of pharmaceutical and biotechnology firms have led to a blurring of the distinction between the two.

THE FIVE BASIC STEPS OF DRUG DEVELOPMENT

There are two fundamentally different types of research conducted in early-stage drug development: basic and applied research (see Figure 4.3). Basic research is directed at improving fundamental knowledge of biological systems, disease processes, and potential points of therapeutic intervention. Applied research utilizes this knowledge to identify and develop the therapeutic agents themselves. This distinction is important because different toolsets and mindsets are involved in these two types of research. Basic research generally does not directly produce drugs. Instead, it lays the foundations upon which drugs are developed and produced. Genomics, proteomics, molecular physiology, and other basic research areas contribute essential information for disease characterization and target identification. Applied research is required to enable further development. Translational research forms a bridge between basic and applied research, testing principles from basic research and generating "proof-of-principle" in preparation for applied research.

Figure 4.4 *The process of drug development*

STEP 1: IDENTIFY A USEFUL DISEASE TARGET

A fundamental understanding of the science involved in a therapeutic problem is essential for drug development. Ideally, scientists start with an understanding of the molecular processes affecting conditions they wish to treat and test hypotheses about which drugs are likely to be effective for a condition.

Model systems are employed because it is unethical and impractical to test uncharacterized candidate drugs on humans. These model systems generally progress in complexity through the stages of drug development. Early-stage model systems may be as simple as a set of molecules in a test tube (subject to knowledge of disease processes) and later-stage model systems may be live animals with diseases similar to human conditions (subject to availability of relevant disease models).

To work on a problem effectively, the tools must be well-defined. Poor model systems or a lack of precise measures can yield experimental results which are misleading or difficult to interpret (see Box *Poorly defined diseases discourage drug development*). It is vitally important to have a fundamental appreciation of the problem to be solved and to apply the appropriate tools to assess potential solutions. Many biotechnology applications, such as gene therapy and RNA interference (described in Chapter 6) are challenged by poor availability of predictive models.

STEP 2: FIND AND REFINE A LEAD COMPOUND

Potential lead compounds typically originate from one of two sources: purified naturally occurring compounds or *de novo* design and synthesis of new compounds. As described earlier in this chapter, biotechnology drugs have different properties and capabilities than traditional pharmaceutical drugs. The traditional pharmaceutical method for drug discovery involves screening libraries of natural or synthetic compounds to find those that achieve the desired effect in model systems. The molecular biology techniques used in

Box
Poorly defined diseases discourage drug development

A significant impediment to developing and testing drugs is the challenge of testing drugs in the early stages of development. Consider the examples of poorly-defined diseases such as Alzheimer's disease, and better-defined diseases such as cancers.

Physical changes in the brain are implicated in Alzheimer's disease, but the precise molecular mechanisms behind these changes are not known. Curing the disease is more complex than simple elimination of these physical changes, because it is not known if these changes are the cause, or a result, of the disease. Further complicating the situation, humans are the only animal to suffer from Alzheimer's, limiting the utility of animal models.

Conversely, many cancers result from defined biological processes. Blocking these processes in *in vitro* experiments and in animal models gives researchers a way to eliminate ineffective drugs prior to engaging in expensive and difficult human trials. Lacking good molecular or animal models, researchers rely extensively on human testing to evaluate Alzheimer's drug leads. Drug testing involves giving patients potentially ineffective or harmful medications and assessing slowed disease progression (or potentially disease reversal) by measuring mental abilities using cognitive-function tests.

The requirement to send relatively more drug candidates to human testing in developing treatments for diseases such as Alzheimer's disease makes drug development more difficult, expensive, and dangerous. Having a strong biomarker, such as an overexpressed gene, or high cholesterol levels, or having a robust animal model, can greatly facilitate drug development. The relative challenge of drug development ultimately impacts how aggressively drug developers will pursue a treatment. If development costs are likely to be high, companies may dedicate their resources to other projects.

biotechnology drug development permit directed selection of natural compounds—usually biological molecules—and the design and synthesis of novel biological compounds as drug candidates.

As knowledge of biological systems improves, it becomes increasingly possible to refine screening methods. Knowledge of disease mechanisms can also help build better model systems through identification of appropriate therapeutic targets as well as other targets which might be a source of undesirable side effects.

An understanding of how biological systems operate at the molecular level, combined with the ability to express gene products in bacteria, yeast, and animal-derived cells, enables development of drugs based on specific disease requirements. Many traditional pharmaceutical drugs were discovered by screening libraries for effective leads, rather than starting with knowledge disease processes and working backwards to find a solution. Knowledge of the structure of a key molecule involved in a disease, such as the HIV protease that is integral to AIDS, enables *in silico* (computer model-based) techniques, using computers to select or design compounds likely to inhibit the enzyme.

Studying herbal remedies used by different cultures, indigenous peoples, and animals is also an excellent source for potential drug compounds. Some naturally occurring compounds, such as penicillin, are used as drugs based on their natural activities in biological systems. The antibiotic penicillin was identified as the factor that permits *Penicillium* mold to inhibit bacterial growth. Whereas penicillin is naturally produced by *Penicillium* as an antibiotic, the same therapeutic application it is used for in humans, other natural compounds, such as *botulinum* toxin, are used for novel purposes such as treating abnormal muscle contractions and in cosmetic applications.

Another method to discover new drugs is to examine the side effects of existing drugs. Minoxidil, now prescribed as a topical treatment for hair loss, was initially intended for the treatment of severe blood pressure. The curious side effect of stimulating hair growth led to its use as a treatment for balding. In another example, Viagra's potential for the treatment of erectile dysfunction was discovered in clinical trials for treatment of angina.

Drug development stages

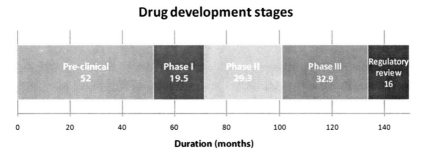

Figure 4.5 *Biotechnology drug development time*
Source: Dimasi, J.A., Grabowski, H.G. The cost of biopharmaceutical R&D: Is biotech
different? *Managerial and decision economics*, 2007. 28:469-479.

Once a potential drug that works in a model system is identified, it is time to study and refine its activity. This potential drug is called a lead compound. Aside from drug activity, factors such as a drug's shelf life at different temperatures, ease of large-scale manufacture, and lot-to-lot consistency must also be considered. In optimizing lead compounds, researchers aim to identify the elements that are essential for their activity and modify those elements to obtain optimal efficacy and/or safety properties.

STEP 3: TEST LEAD IN PRE-CLINICAL DEVELOPMENT

In pre-clinical development, lead compounds that emerge from the lead optimization process are subjected to a range of standardized animal, cellular, and biochemical tests designed to gauge their suitability and safety for human administration as well as to estimate the range of dose levels of the compound that will be utilized in subsequent human trials. Animal models are also used to provide preliminary assessments of the absorption, degradation, and potential toxicity of drugs. During pre-clinical development, these animal tests are conducted under much more tightly prescribed conditions, such as the industry-standard GLP (Good Laboratory Practices) procedures, which have stringent quality control and quality assurance oversight than is normally used in earlier stages. While it may be possible to produce a desired effect in a model system in a laboratory setting, real-world situations often present unforeseen obstacles. For example, many gene therapy techniques that work in cultured cells in laboratory settings fail when introduced into human beings.

Scientists use animals to test toxicity and attempt to cure animal versions of human diseases before proceeding to human trials. Because biotechnology enables biological changes that were previously impossible, it is not possible to predict all the implications. Many products that work well in laboratory tests fail in clinical settings for unforeseen or even improbable reasons; drugs may not be taken up properly by cells; they may be metabolized into inactive or toxic forms by the liver; they may interact with other parts of the body to produce undesired effects; or they may simply not be sufficiently active. This is one of the reasons why animals are so important in drug research.

While success in animal tests does not necessarily mean that a compound will work in humans, a compound that performs poorly in animals is unlikely to work well in humans. Ultimately, human testing is necessary for the safety and efficacy determinations required for FDA approval.

In addition to testing the drug candidate in animal models, another set of activities that takes place during pre-clinical development is development of methods for manufacturing and formulating the drug on a commercial scale. Unlike research costs, manufacturing costs recur over the life of a drug. Minimizing these recurring costs can significantly impact profits. This examination is also important for patent protection because it may lead to additional patent claims, potentially impeding the development of competing drugs.

STEP 4: CLINICAL TRIALS IN HUMANS

The clinical trial process, described in Chapter 8, investigates drugs for safety and efficacy in humans. There are four "phases" of clinical trials. Phases I through III demonstrate safety and efficacy prior to approval, and Phase IV monitors safety post-approval and tests new treatment indications.

Briefly, drugs are first tested in a small group of healthy, or in some cases affected, individuals to determine safe dosage limits. Larger trials follow to investigate safety in diverse populations, establish dosage regimens, and demonstrate efficacy. Drugs adminis-

Biotechnology drug approval times

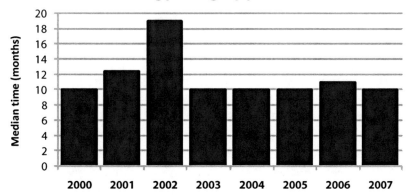

Figure 4.6 *Median approval times for biotechnology drugs*
Source: Recombinant Capital

tered in clinical trials must be produced using current good manufacturing practices (cGMP), ensuring proper control of facilities, raw materials handling, manufacturing, and associated documentation. Clinical trial data is submitted to the FDA as part of a New Drug Application (NDA) or Biologics License Application (BLA).

STEP 5: OBTAIN APPROVAL; MARKET AND SELL DRUG

The FDA requires that drugs be approved prior to marketing. While safety is the primary concern, a drug with detrimental side effects may be acceptable if there are no better treatments and the severity of the disease warrants it.

Current estimates of development times for small-molecule drugs are 10-15 years with an estimated average cost of $802 million per approved drug.[1] It is worth noting that half of this cost is attributed to financing costs, reflecting the "opportunity-cost of capital" invested over the 10-15 year timeline. Roughly one-third of the expenditures are attributed to pre-clinical activities and the remaining two-thirds to clinical activities (for further discussion of the cost of drug development, see Box *The cost of drug development* in Chapter 12). The Tufts Center for the Study of Drug Development found that only five in five thousand small-molecule compounds that en-

---

1 DiMasi, J.A., Hansen, R.W., Grabowski, H.G. The price of innovation: new estimates of drug development costs. *Journal of Health Economics*, 2003. 22:151–185.

ter pre-clinical testing make it to human testing. Of these five, only one is approved.[2] These numbers were derived from examination of small molecule synthetic drugs, which are produced by traditional pharmaceutical techniques and differ from biologic drugs in several important ways. Estimates for the cost of biologic drug development are $1.2 billion, comprised of approximately $500 million in out-of-pocket expenses, and $700 million in capitalization costs.[3] Both these cost estimates include the cost of failed leads. Therefore, they do not predict the expenditures required to produce a single drug; they predict the investments required by successful and failed research projects that result in the development of a single drug.

Relative to small molecule drugs, the sample size to evaluate development times and costs for biologics is much smaller and subject

2   How new drugs move through the development and approval process. *Tufts Center for the Study of Drug Development*, November 1, 2001.

3   Kaitin, K.I. (ed.) Cost to develop new biotech products is estimated to average $1.2 billion. *Tufts Center for the Study of Drug Development Impact Report*, 2006. Vol. 8.

Box
The Human Genome Project and drug development

The Human Genome Project (HGP) was a multi-billion dollar multinational effort to sequence the entirety of the human genome, identify all the genes, improve tools for genome analysis, and address related ethical, legal, and social issues. The project started in 1990 and sequencing was completed ahead of schedule in 2003.

The initial findings from the HGP informed scientists that molecular biology was far more complicated than previously believed. Projections for the number of genes in the genome, for example, ranged from the high tens-of-thousands to more than 100,000. After examining the sequence of the human genome, it was found that the genome contained far fewer genes than previously suspected—less than 30,000. The key to enabling the genome to produce a sufficient diversity of proteins from this relatively small set of genes is in editing individual gene mRNAs so that each gene can produce multiple proteins. Entirely new methods for regulation of gene expression were also discovered, further complicating efforts to tame molecular biology.

So, why has the HGP not yet produced a revolution in drug development?

- The HGP was primarily a basic science endeavor. It has greatly expanded the knowledge-base essential for drug development
- Information gleaned from the HGP must be interpreted and understood before it can produce new drug leads
- It can take more than a decade for a drug lead to gain regulatory approval
- Most drug leads fail to gain regulatory approval

While processing the new information from the HGP will occupy researchers for decades to come, there are some immediate benefits. The project brought many technological advances. The costs of synthesizing and sequencing DNA, for example, decreased by several orders of magnitude and the precision of many experiments has improved, along with the ability to automate many procedures. These improvements have translated beyond humans into agriculture, where farmers are better equipped to identify top-performing plants and select them for traditional breeding programs. DNA fingerprinting, which has greatly advanced the field of forensics, is also a spin-off of the human genome project. These advances and others are benefitting science today, while we await the other outcomes of the HGP.

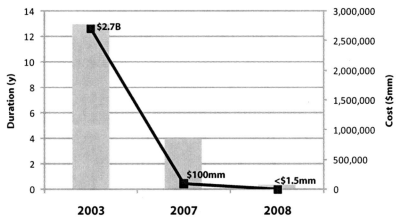

Figure 4.7 *Declining time and cost of human genome sequencing*
Source: James Watson's genome sequenced at high speed. *Nature*, 2008. 452:788.

to bias—the first biologics had shorter development times than later entrants—but initial indications are that development times and costs for biologics are similar to those for small molecules. It is important to note that any estimate of drug development time or cost is profoundly affected by context. First-in-class drugs, drugs serving new markets, or drugs serving pressing needs are likely to require smaller and fewer clinical trials than drugs with little differentiation from existing alternatives or those serving less-pressing needs, decreasing the time and cost of development.

Once a drug receives regulatory clearance for marketing, it will likely be protected by patents that were filed before clinical trials began. With an average of twelve years of patent protection remaining after FDA approval, marketing and sales efforts must generate revenues and expand market penetration to deliver a return on R&D expenditures. After a drug is on the market, drug sponsors must monitor patients for unexpected side effects. Independent or sponsored clinical trials can also test suitability for additional indications, potentially expanding the market. The emergence of competing products and looming patent expiration dates motivate the development of alternative drug forms and formulations to leverage established brands, and modification of marketing methods to extend sales. These strategies are covered in greater detail in Chapter 12. Some firms specialize in modifying patented drugs, capitalizing on the negative specter of patent expirations, and patent their modifications to license them to pioneers as a means to preempt generics.

The aforementioned five steps of drug development must always occur, and the influences of individual biotechnology innovations are compartmentalized. Each innovation can only affect one or a few steps. For example, functional genomics can aid in developing model systems and selecting or designing potential drugs, but cannot resolve drug delivery or manufacturing issues. Molecular evolution can refine lead compounds, but cannot assist discovery, clinical trials, or manufacturing. The impact of this compartmentalization is that no single technology can profoundly alter the process of drug development; a number of complementary innovations are necessary for a revolution.

Table 4.1 *Qualities of a "good" drug*

Market quality	Pharmacological qualities	Barriers to entry
• Market size • Dosage frequency • Manufacturing cost • Price elasticity	• Safety • Effectiveness • Chemical stability • Metabolic stability • Drug delivery	• Intellectual property protection • Market exclusivity • Challenge of generic manufacture

QUALITIES OF A "GOOD" DRUG

Beyond safety and efficacy, many other factors influence the quality of a drug. An ideal drug must address a market that is willing to pay a price that permits profitable sales. The commercial attractiveness of a drug is influenced by the size of a drug's patient population, the frequency of dosage, cost of production, barriers to entry of competitors, and availability and cost of alternative treatments.

Consider the example of penicillin, the first antibiotic. The devastating effects of bacterial infections and the absence of effective treatments guaranteed strong sales. The raw materials for penicillin production were known and initial tests showed a good safety profile. The remaining challenge for penicillin production was developing a method to produce sufficient quantities of the drug at a reasonable price.

In addition to safety and efficacy considerations, practical aspects such as chemical stability and therapeutic administration also affect the commercial prospects of a drug. Oral administration and patches are generally preferred to injections. Unfavorable administration can reduce patient compliance or place a drug at a competitive disadvantage to alternatives. Some drugs must also be mixed with carriers, which can impart their own side effects (see *Drug Delivery* in Chapter 5).

The chemical stability of a drug affects the conditions under which it can be distributed to pharmacists and stored by patients (e.g., Is it heat sensitive? Must it be refrigerated?). The metabolic stability determines how long the product remains effective in a patient's body and how much product must be administered to achieve a therapeutic effect. Metabolic stability also influences how many daily doses are necessary, and it can also be a factor in side effects. A drug's solubility affects numerous outcomes: if it will be stored in fat, whether or not it can travel through the bloodstream, if it can cross the gastrointestinal lining or be delivered through the skin, and if the kidneys can excrete it. These factors impact drug delivery, dosage regimens, and the potential for side effects. Toxicity or side effect limits may be relaxed if the alternatives to drug treatment warrant it.

Copying drugs is a sufficiently lucrative business to motivate companies to specialize in challenging patents and developing generic versions of drugs with expired patents (see *Generic Drugs* in Chapter 8). To prevent competitors from capitalizing on the efforts of pioneers and selling drugs for prices which reflect their reduced R&D burden, drugs must have some form of commercial protection. Drugs may be protected by patents, trade secrets, or other methods such as FDA-granted temporary market exclusivity. The vehicles that offer exclusive market rights for drugs—described in Chapters 7 and 8—are designed to promote innovation by granting drug developers temporary monopolies that permit them to recoup their investments in research and development.

Chapter 5

Tools and Techniques

> I think the evidence is overwhelming that one tool doesn't do
> it. Yet time and time again, we see new entrants coming into
> this business saying, "This tool will revolutionize the discovery
> process," when it's much more likely that the integration of
> tools, of how they work, will have a much more powerful effect.
> *Harvard Business School Professor Gary P. Pisano*

It is only through an understanding of the tools and techniques used for biotechnology research that one can develop an appreciation of the possibilities and challenges of biotechnology. This chapter presents a survey of biotechnology tools and techniques to foster such an appreciation.

The tools and techniques used for biotechnology research define the universe of products and services that biotechnology companies can develop. While a biotechnology company could decide to focus on a set of applications or a set of technologies, most define themselves around technologies rather than applications. From a practical perspective, it is simpler to focus on a technique that can have several applications than to search for all the methods to solve a single problem. Whereas tackling a specific application may require expertise in numerous techniques, a single technique can be used for multiple applications, positioning research and development to serve multiple markets reduces market risks associated with any individual market. Therefore, it is simpler and often preferable to develop expertise and patents for a few techniques and exploit this competitive advantage to develop solutions for multiple applications.

BIOINFORMATICS

Bioinformatics is the convergence of information technology and biotechnology, applying information technology to manage and analyze the vast amounts of data generated from basic biological research. Bioinformatics assists scientists in managing data and enables interpretation of data by presenting it in useful formats.

The tools and techniques that define bioinformatics are themselves a demonstration of the growth and diversity of techniques in drug discovery. As late as the mid-20th century, drug discovery was conducted mainly through chemical synthesis followed by extensive trial-and-error testing. Large-scale testing of derivatives of potential drugs was introduced in the 1970s, followed by attempts at rational drug design in the 1980s. Bioinformatics entered the arena in the 1990s, enabling drug synthesis and testing to be simulated by computers. As biological knowledge, computational power, and computer algorithms improve, it becomes increasingly possible to identify and refine potential leads through the use of computers.

There are two successive elements in bioinformatics: data assembly and data analysis. Computer-assisted data management enables gathering, analysis, and representation of biological information, helping scientists better understand biological processes, understand the mechanisms behind diseases, develop methods to treat diseases, and develop other applications based on biological knowledge. Bioinformatics also allows researchers to perform comparative and predictive studies of biological processes. Applications of bioinformatics data analysis include prediction of protein structure, prediction of protein function, and drug target selection.

One important development that emerged at the beginning of the twenty-first century was the implementation of automated research techniques such as DNA sequencing and robotic fluid-handling and assay systems, permitting large-scale research efforts. A defining event in the automation of biological research was sequencing the human genome. This mammoth project, extending over a decade, determined the sequence of the three billion base-pairs that comprise human DNA. The logistical problems of collecting and managing this mass of information required the development and application of novel computer technologies to assist biological research.

Sequencing the three billion base-pairs of the human genome was only possible with the use of bioinformatics to manage all the data. The information that can be processed using bioinformatics techniques includes not only sequence information for genes and proteins, but also details on the structure and function of proteins, disease correlations, and raw information produced from scientific experiments such as microarray analyses and protein interaction studies.

Bioinformatics applications exist for most steps of drug development. Predictive and analytical algorithms can screen potential lead compounds or help design them from scratch; toxicity can be predicted by comparison against compounds with known properties; even clinical trials can be simulated. Any biological information that can be entered into a computer database is subject to processing and analysis by bioinformatics.

A strength of bioinformatics is the ability to extract information and identify patterns from large databases. Data mining, an analytical bioinformatics application, uses computers to analyze masses of information. Integration and comparison of numerous experimental observations permits the discovery of trends and patterns in large databases, potentially identifying novel relationships.

A significant challenge in bioinformatics is extracting useful knowledge from the masses of information present in databases. Of the many patterns and correlations that may be found, how many are predictive of real-life scenarios? Bioinformatics analysis can make many predictions based on interpretation and extrapolation of data, but the information is only as good as the scientific models and data used to derive the predictions. Traditional experiments are still required to confirm and verify bioinformatics-derived predictions and simulations.

Another challenge for commercial bioinformatics companies is securing a competitive advantage to permit information and tools to be sold at a profit. Unlike defined products such as drugs and research platforms that can be effectively protected by patents, biological information and computer programs are more difficult to protect. The case of *Bayer v. Housey*, described in the Box *When research is done abroad* in Chapter 16, describes how it may be possible for companies to use patented drug screening technologies abroad and import the results without penalty. Competing companies, academic researchers, public initiatives, and open-source projects are also free to develop competing bioinformatics tools or publish research findings that can undermine proprietary databases. Furthermore, whereas traditional research products such as laboratory equipment may endure for years without modification, increases in computing power and changes in scientific paradigms can lead to bioinformatics tool obsolescence in a matter of months, requiring constant R&D.

CLINICAL MODELING

Clinical trials are necessary to demonstrate that drugs are safe and effective. Drug developers are not permitted to market drugs until they receive FDA approval, based on safety and efficacy data generated in clinical trials. Time spent in clinical trials is time that cannot be spent selling drugs. Because drug patents must be filed prior to the initiation of clinical trials, any reduction in clinical trial duration can be valuable—a single day's delay in approving a billion dollar blockbuster drug can mean a loss of revenue exceeding $3 million.

The processes for clinical trials are flexible—they can be adapted for diverse drugs—positioning clinical research organizations to build businesses around this critical step in drug development. By focusing on clinical trials, contract research organizations are able to develop specialized expertise and relationships with clinical trial providers. Clinical research organizations generally offer two distinct solutions to aid clinical trials. Trial management solutions involve strategies to design trials, facilitate patient recruitment, speed and improve communication between trial investigators and sponsors, and manage and analyze data. Simulation and prediction services use sophisticated computer techniques to enable the safety and efficacy of drugs to be predicted at a lower cost and with greater speed than actual clinical trials, ultimately permitting selection of compounds and trial protocols most likely to lead to FDA approval.

COMBINATORIAL CHEMISTRY

Combinatorial chemistry is a general method for creating a large number of molecules and systematically testing them for desired properties. An example of the automation of research, combinatorial chemistry is roughly analogous to a structured implementation of setting one million monkeys typing randomly on typewriters in the hope that at least one will produce a complete novel.

Combinatorial chemistry systems are designed to produce a variety of molecules according to a set of predefined rules. The library of synthesized molecules is then tested via high-throughput screening in a model system, possibly an enzyme-activity or protein-binding assay, to determine which compounds show promise for a given application. A combinatorial chemistry experiment may demonstrate interaction between two compounds or the ability of test compounds to enable or inhibit a chemical or biological reaction. Promising compounds may be used for further rounds of combinatorial chemistry, to search for related compounds with improved properties, or may be used directly as target compounds for further development and testing.

Variations on combinatorial chemistry such as combinatorial genetics apply the same general paradigm to different kinds of problems. Combinatorial genetics, a form of molecular evolution, involves the mutation of a gene to produce a library of variants. These variants are then analyzed for desired qualities. This technique has particular appeal for industrial biotechnology, where it can aid in the search for improved enzymes with improved yields, stability, or other characteristics.

The commercial potential of combinatorial chemistry is limited by low barriers to entry. Combinatorial chemistry companies must offer a competitive incentive for customers to outsource rather than develop abilities internally. Furthermore, individual companies are challenged to prevent commoditization by offering unique value propositions and differentiating their offerings from competitors.

FUNCTIONAL GENOMICS

Functional genomics seeks to understand the activities of genes in healthy and diseased states. The reality of human genetic variation means that different patients respond differently to the same drug. The difference may be as simple as a slight difference in efficacy or it may result in a drug being completely ineffective or even toxic in some patients. It is estimated that most commonly used drugs are effective in only 30–60 percent of patients with a given disease. A subset of these patients may suffer severe side effects.

Without functional genomics there is no simple way to determine if a given patient or subset of the population is likely to respond either well or poorly to a medication. As a result, drugs are developed for the "average patient." Furthermore, many drugs that might benefit a subset of patients may never be developed because they cannot be shown to be useful in an average group of patients. Patients who are unlikely to benefit from a drug or who may suffer adverse side effects are likewise not identified and given more appropriate treatments. Functional genomics enables segmentation of patient groups to resolve these issues.

Knowledge of the sequence of the human genome is a valuable tool for functional genomics. However, simply knowing the sequences of genes is not sufficient. Discrete genetic differences between individuals must be correlated with the effects of medications. Studying single nucleotide polymorphisms (SNPs) and pharmacogenetics reveals correlations that enable functional genomics.

SNPs are discrete DNA sequence changes between individuals that are at the root of many genetic differences. SNPs have been linked to the likelihood that an individual will find a drug effective or unsafe. These therapeutic differences are related to variations in drug targets, in enzymes that metabolize drugs, and in other molecules involved in cellular metabolism. The elucidation of discrete genetic differences that can be readily identified holds the potential to predetermine how a patient will respond to a drug.

PHARMACOGENETICS AND PHARMACOGENOMICS

Pharmacogenetics studies the relation between genetic variation and the effects of pharmaceuticals: the investigation of how genetic differences affect the ways in which people respond to drugs. Specifically, pharmacogenetics seeks to understand the differences between drug targets and metabolic enzymes that affect efficacy and toxicity. Differences in genetic sequences are responsible for many of the differences between individuals. Just as genes influence eye color and hair color, they can also influence susceptibility to disease and determine whether specific drugs are safe and effective for certain individuals. The terms pharmacogenetics and pharmacogenomics are sometimes used to respectively distinguish between the correlation of single drugs with multiple genomes, and of multiple drugs with single genomes.

Learning why certain individuals are unresponsive to drugs or experience dangerous side effects gives researchers the potential to develop drugs that address these shortcomings. Studying the mechanisms by which drugs are rendered ineffective or toxic may also enable drugs to be designed to avoid or compensate for these alterations. Furthermore, drug discovery cost and time can be reduced by eliminating potential clinical trial participants for whom drugs in development are likely to prove ineffectual. More precise clinical trials justify smaller and fewer trials, facilitating FDA approval (see commentary on Herceptin clinical trials in Box *Personalized medicine and drug sales* in Chapter 6).

Functional genomics can facilitate drug discovery and improve drug administration. Applying functional genomics in a personalized approach to medicine—prescribing drugs only to patients likely to benefit from them—can streamline medical care and avoid unnecessary side effects. Functional genomics can also potentially identify patient groups of less than 200,000 Americans, qualifying drugs for Orphan Drug status (see *Orphan Drugs* in Chapter 8).

Specific challenges to implementing functional genomics are the development of tools to profile individual patients and retooling drug development for smaller patient groups. In order to prescribe drugs appropriately to an individual's genetic profile, a system to rapidly

Box

Cytochrome p450 and pharmacogenomics

Cytochrome p450 is a generic term for a set of enzymes which are collectively the most important element in chemical modification and degradation of chemicals including drugs and other foreign compounds. A vast majority of the most serious adverse reactions to medicines appear to involve drugs that are metabolized by the cytochrome p450 system.[1]

Six different p450 genes are responsible for most of the metabolism of commonly used drugs. Each gene can have dozens of discrete mutations affecting its activity. Inventorying the set of cytochrome p450 enzymes and elucidating the factors contributing to their expression and activity levels is central to understanding and predicting differences in response to drugs. Some of the drugs and compounds degraded by cytochrome p450 enzymes are caffeine, morphine, Taxol, Prilosec, cocaine, codeine, Viagra, St. John's wort, and HIV protease inhibitors. If two compounds are degraded by the same cytochrome p450 enzyme it is possible that taking both compounds at the same time can lead one or both to accumulate to dangerous levels. This is one of the ways in which drugs can interact to alter efficacy or have lethal consequences.

Roche's AmpliChip 450 is the first microarray-based diagnostic test that can detect genetic variations influencing drug efficacy and adverse drug reactions. The AmpliChip contains 15,000 DNA sequences representing 31 genetic variations in two cytochrome p450 enzymes. According to Roche, the two enzymes affect 25 percent of commonly prescribed medications. The purpose of the chip is to determine whether a patient metabolizes drugs at a normal, slow, or fast rate. This information can help doctors prescribe appropriate medications and dosages based on a patient's rate of degrading specific drugs. Properly calibrated dosages can mean the difference between no response, therapeutic effectiveness, and serious side effects.

---

1 For a topical review, see: David A. Katz, D.A., Murray, B., Bhathena, A., Sahelijo, L. Defining drug disposition determinants: a pharmacogenetic–pharmacokinetic strategy. *Nature Reviews Drug Discovery*, 2008. 7:293-305.

and inexpensively determine appropriate elements of an individual patient's genetic profile is necessary. The gains in safety and efficacy of drugs with functional genomics-based prescription are coupled with smaller patient populations. Accordingly, drug firms must be willing to actively exclude patients in order to realize the benefits of functional genomics. Smaller patient populations don't necessarily mean smaller profits—see the Box *Genzyme: Building an empire on orphans* in Chapter 8.

MICROARRAYS

Microarrays are tools that permit the identification of DNA or other samples and examination of gene expression and protein modifications in individual tissues, and under different conditions. While the first microarrays were directed at detecting DNA, new technologies have enabled protein-based and other forms of microarrays. Microarrays enable researchers to detect the presence or expression of many genes and proteins at once. The ability to simultaneously examine the changes in expression of many different genes, or changes in protein levels and modifications, is useful in investigating the effects of diseases, environmental factors, drugs, and other treatments in human health.

Applications of microarrays include diagnosing or identifying cancerous cells, assessing genetic predispositions to diseases, examination of gene expression, and gene and protein responses to drugs or other therapeutic procedures.

The ability of microarrays to convey information on the status of thousands of genes and proteins leads to data management challenges. Sophisticated computer techniques and methodologies are required to efficiently gather information from microarray experiments and enable interpretation.

PROTEOMICS

Proteomics is the study of protein structure and function. Genes encode proteins, which perform structural and functional roles in cells. It is proteins, not genes, which are the major actors in molecular biology (see Chapter 3). Understanding the structure and function of proteins can lead to new therapies and influence disease diagnosis and treatment.

Proteins can be roughly categorized by their structural and functional roles. An example of a structural protein is keratin, a component of skin. Functional proteins, called enzymes, perform cellular duties. Metabolic enzymes aid in food digestion and enable harvesting of stored energy. Studying DNA can reveal some information on the control of protein synthesis, but provides limited information about the structure and function of proteins. This requires examination of the proteins themselves.

Proteomics uses a variety of techniques to examine protein structures and functions. Unlike DNA sequencing, which is a relatively uniform technique that can be widely used without modification, the very methods used to investigate proteins vary with each individual protein being studied. There is no simple or uniform way to produce, identify, quantify, or characterize proteins. The need to continually adjust experimental methods in proteomics research is a significant challenge to scaling or automating research efforts.

An additional challenge for firms providing proteomics services is the need for profit-enabling business models. Just as the distance between genomics and drug approval challenges companies selling genomic information databases to demonstrate the value of their wares, proteomics information providers also need to prove their worth. Proteomics companies may offer potential lead compounds or drug targets, but much time and expense is still required to develop approvable drugs from this information. The value of proteomics research aids—protein arrays, interaction maps, and biological assays—are also difficult to assess objectively, making it difficult to command high prices. Despite the challenges of commercializing proteomics, research in this area holds the potential to increase knowledge of biological systems and enhance biotechnology product research and development.

MANUFACTURING

Developing a drug that is safe and effective is essential to gain FDA approval, but to generate revenues and recoup development costs it is necessary to manufacture and sell the product as well. Conventional wisdom once held that investing in manufacturing process development did not benefit a company's returns as much as basic research. In an era of increased competition where companies frequently produce competing treatments, the ability to accurately predict demand, to rapidly develop production methods, and to scale production capacity provide a strategic advantage (see Box *Enbrel: Underestimating market demand* in Chapter 13).

In the process of drug development a drug may be produced in test tubes, flasks, small fermentation vessels, pilot-plants, and large-scale production facilities. The progression from bench-top production to large-scale production is not a trivial process. As the scale of production increases, factors such as temperature, oxidation, and mixing change, potentially altering the final product. To ensure drug quality, manufacturers must demonstrate compliance with FDA current good manufacturing practices (cGMP) and further prove that drugs of consistent purity and activity can be produced in large quantities from batch to batch, day after day, year after year.

Manufacturing facilities for purified biological products have special requirements. Sophisticated equipment is required to control environmental conditions and workers must be trained to manage the facilities. Ensuring purity requires air, water, and reagent purification and sterilization systems. Product consistency further demands that water, nutrients, and other reagents are pure and of consistent quality. Semiconductor fabrication and drug manufacturing reportedly require more water than any other high technology applications. It is estimated that as much as 30,000 kilograms of water are required for every kilogram of protein drug. This water supply must be uninterrupted and of high quality. Regional water rationing or failure of municipal facilities to increase water capacities with product demand may limit national or international product supplies.

One alternative to traditional manufacturing methods is the use of animals and plants that are genetically modified to produce a desired compound. For example, drugs may be harvested from chicken eggs or cow milk, or purified from plant tissues. A significant benefit of transgenic production is that production of raw material is relatively simple to implement and scale. It is estimated that plant-based

Box

GMP: Building quality into products

Biotechnology companies intending to sell products internationally need to address local regulatory requirements of the targeted markets and to ensure that the products are pure, safe and effective for the consumers. The laws may be different in different countries, but the objectives are the same: to ensure human and veterinary safety. In assessing facilities, auditors look for compliance to cGMP (current Good Manufacturing Practices).

GMP is part of quality assurance which ensures that products are consistently produced and controlled to the quality standards appropriate to the intended use. GMP builds quality into products through the following key elements:

- Personnel
- Contamination control
- Premises
- Equipment
- Defined processes and procedures
- Documentation and records
- Change control
- Effective use of validation

It is very important to recognize that quality cannot be "tested into" products. Consider the following quality control protocol for an injectable:

Lyophilized Parenteral Product
Batch size: 50,000 cartridges
Sample size for testing: 100
Tests performed: Description, ID, assay, pH, endotoxin, sterility, particulate analysis, moisture

Unknown or not tested parameters include grease, cross-contamination (e.g., penicillins), extraneous matters, etc. The only way to achieve a high quality of product is to start at the pre-commercial stage: have trained personnel, build the right facilities (e.g., choose the cleanroom standards according to product requirements), practice regular cleaning and sanitization, document diligently, review and control any changes, perform effective validation, etc.

Once a licence is obtained, continued control, vigilance and maintenance of the system is needed for the licence to be renewed. If serious non-compliance is observed, the manufacturer is liable for fines, prosecution, withdrawal of products, and/or potential facility shutdown.

Contributed by Poh Kian Toung (kiantoungpoh@yahoo.co.uk), Manufacturing Quality Manager.

biologic production can be 10 to 1000 times less expensive than conventional fermentation systems. Leveraging the relative simplicity and cost advantage, drug-producing varieties of animals and plants can be distributed in regions lacking sufficient expertise, facilities, or resources for conventional fermentation production. To scale production, these transgenic factories can be bred or cloned, increasing the supply of raw product.

A drawback of transgenic production is the long development time relative to conventional methods such as microbial or cellular fermentation. While scaling from pilot to commercial-scale production is simply a matter of farming more transgenic animals or plants, the initial development is more time consuming. In competitive environments, speed to clinical trials may be more important than speed to commercial production, because announcing positive clinical trial results earlier than competitors can help companies attract more financing, ultimately granting them more flexibility for downstream activities like manufacturing. Therefore, when the focus is on clinical trials, transgenic animal and plant-based methods must compete with traditional fermentation. The ability to delay the bulk of production development until after a drug has shown promise in clinical trials favors conventional production, permitting companies to invest only in drugs that are likely to be approved.

The cost savings of transgenic plant and animal-based production are also offset by the need to build protected and contained growth facilities to prevent accidental environmental release and mixing with wild populations.

DRUG DELIVERY

Despite the emphasis on the biological activity of drugs, it is important to also consider the systems and methods used to deliver drugs to their therapeutic targets. The goal of drug delivery systems is to enable active medications to reach appropriate parts of the body, in the appropriate concentrations for the appropriate amount of time, where they can accomplish their therapeutic task.

Advanced drug delivery systems, which can improve some traditional pharmaceuticals, are essential for many biotechnology drugs. The majority of traditional pharmaceutical compounds are small molecules that are administered orally as pills or liquids and reach their therapeutic target via the bloodstream. Advanced formulations of traditional pharmaceuticals enable control of factors such as duration and rate of delivery, or targeting to specific tissues.

Many side effects are caused by the carriers needed to deliver drugs, rather than the drugs themselves. For example, castor oil, which is used to make Taxol sufficiently soluble for injection, is associated with numerous side effects such as nausea, vomiting, joint pain, appetite loss, brittle hair, and tingling sensations in hands and feet. Furthermore, the risk of serious allergic reactions necessitates prior administration of steroids and antihistamines. In such cases, the development of improved delivery methods is essential to realize the full benefit of otherwise potent therapeutics.

Drugs produced by biotechnology techniques tend to be large proteins and nucleic acids which face special challenges relative to smaller, more chemically stable pharmaceutical drugs. Factors impeding oral delivery of biologic drugs include:

- Acidity of the digestive system
- Intestinal enzymes that degrade proteins
- Inability of biologic drugs to cross intestinal walls
- Poor solubility of biologic drugs

Overcoming the challenges of biologic delivery can present new opportunities, as targeted and metered dosage systems can potentially improve drug effectiveness, mitigate safety and side effect con-

Table 5.1 *Selected nanotechnology applications in drug delivery*

Technology	Benefit
Carriers	Improve solubility and avoid need for harsh solvents
Encapsulation	Extend duration of drug bioavailability
Nanoparticles	Improve solubility, speed delivery, and extend duration of drug bioavailability

cerns, and ultimately improve patient compliance and retention.

Patient compliance is also a concern in drug delivery. The requirement to take many pills a day, to follow rigorous dosage regimens, or the use of unappealing delivery methods such as injection may discourage compliance, and patients cannot benefit from drugs they don't take. Reducing administration from several times a day to once a week by using an extended-release formulation, for example, can dramatically improve compliance and ultimately improve patient outcomes—the primary objective of drug therapy.

Delivery techniques that increase compliance can ultimately help a company derive more revenue from a product. Selling twice as much of a drug by doubling the duration that patients take the drug or doubling the number of people who take it is arguably similar to selling two drugs, without the cost of developing two drugs.

NANOTECHNOLOGY

Nanotechnology is a multidisciplinary field encompassing the development and application of materials at sizes measured in billionths of a meter. Surface tension, molecular interactions, and surface area exposure play an increasingly important role in chemical and physical interactions at this size range, giving nano-scale materials properties that are markedly different than those seen at larger scales. Nano-sized flour particles, for example, are capable of igniting violently and causing explosions in flour mills. Geckos are able to climb walls because of nano-scale hairs on their feet which use atomic interactions, rather than stickiness, to adhere. The enzymes and other key molecular players in biotechnology also operate at this size scale, creating an opportunity for convergence between biotechnology and nanotechnology.

As with biotechnology, early investments in nanotechnology research were attracted by the high profit potentials of serving unmet medical needs. Beyond developing new therapeutic products with nanotechnology, there are also strong opportunities in developing delivery systems that can improve the safety and efficacy of existing drugs (see *Drug Delivery* earlier in this chapter). Reducing the particle size of drugs has the potential to:

- Increase surface area
- Enhance solubility
- Improve oral bioavailability
- Speed onset of therapeutic effect
- Decrease necessary dosage
- Decrease variability between fed and fasted dosage
- Decrease patient-to-patient variability

Elan Pharmaceuticals' NanoCrystal technology overcomes solubility problems by using a proprietary technique to absorb nanoscaled particles of drug substance onto the surface of stabilizers. NanoCrystal technology allows drugs to be released into the bloodstream faster than alternative methods and enables drugs to stay dissolved for longer periods of time. In August 2000 Wyeth's solid-dose form of the immunosuppressant Rapamune became the first FDA-approved drug incorporating NanoCrystal technology. Rapamune was previously available only as an oral solution, requiring refrigerated storage and mixing with water or orange juice prior to administration. NanoCrystal tablets enable a solid dose formulation which does not require refrigeration.

Abraxane is an example of how nanotechnology can improve existing drugs. It was developed as an improved version of Taxol. The castor oil carrier required to solubilize Taxol (see *Drug Delivery* earlier in this chapter) has several significant side effects associated with it. By binding paclitaxel (a generic version of Taxol) to nano-scale protein particles, Abraxis Oncology produced a version of the drug that could be injected without castor oil, dramatically improving the side effect profile.

Nanotechnology also has many applications in research. Researchers continually seek to perform assays using smaller and smaller quantities of experimental materials, which are often very expensive and difficult to obtain. Smaller-scale experiments provide two advantages: the ability to perform more experiments at lower costs, and the ability to run more experiments in less space. Microarrays, for example, enable researchers to examine many genes at once using a small quantity of material. Nanotechnology-based solutions, such as "lab on a chip" products, can reduce the scale of experiments even further, enable multiple sequential or simultaneous assays, and expand opportunities for automation in research.

Chapter 6

Applications

The best way to predict the future is to invent it.
Richard Feynman

Biotechnology companies focus on selling products or offering services. Products include drugs, reagents, research tools, industrial enzymes, and specialized crop plants. Services include discovering drug lead compounds, clinical trial management, and manufacturing. More details on biotechnology business models are provided in Chapter 10.

The most common application for biotechnology companies is drug development. This is partially due to the enormous profit potential of drugs, which can greatly offset the increased development cost relative to other applications. Some firms specialize in elements of the drug development process, supporting the search for potential drug compounds by coordinating clinical trials or producing research tools and drug delivery systems. These "pick and shovel" companies benefit indirectly from the profits of other development companies by selling necessary products and services. While some service firms fulfill functions that could be developed internally, dedicated service firms may also possess economies of scale, enabling them to offer specialized expertise and abilities.

Drug development differs from most other commercial development ventures because products must be proven safe and effective before they can be marketed. Applications not intended for human use benefit from not requiring clinical trials to prove their safety and efficacy prior to commercial release, although these applications may still be subject to EPA, USDA, and other political or ethical re-

Table 6.1 *Biotechnology application categories*

Category	Description
Green: Agricultural biotechnology	Products and applications related to livestock and crop production, and agricultural production of biotechnology products.
White: Industrial biotechnology	Modification or improvement of industrial processes such as paper processing, bioremediation, and chemical and organic compound synthesis.
Red: Medical biotechnology	Drugs and other agents to treat, cure, or prevent disease, and products that assist in the diagnosis of diseases or measurement of crucial factors in health and disease.

strictions. While clinical trials are often seen as a hurdle, Figure 12.1 in Chapter 12 and the adjoining box *Clinical trials provide valuation milestones* explain how clinical trials can be an asset in obtaining funding. The regulatory controls influencing biotechnology are discussed in detail in Chapter 8.

Table 6.1 shows the division of biotechnology applications into three categories: Green biotechnology for agricultural applications, white biotechnology for industrial applications, and red biotechnology for therapeutic applications. Specific applications are described in further detail in this chapter.

In reading this chapter, it is important to recognize that biotechnology is not a panacea. In addition to practical considerations such as technological constraints, the legal, regulatory, political, and commercial factors described later in this book have profound impacts on the ability to develop and commercialize biotechnology. Many biotechnology companies fail because they develop products for which profit-enabling markets do not exist.

GREEN BIOTECHNOLOGY: AGRICULTURE

The directed modification of plants and animals can increase their value in agricultural applications. By studying the genes responsible for specific traits, it becomes possible to introduce, alter, or change the expression of those genes in a controlled manner, resulting in a desired change. Extensive testing, mandated by the FDA, EPA, and USDA, is required to determine if genetically modified plants are safe for humans and ensure that they do not pose a threat to the environment. This testing requires demonstration that foreign proteins in edible crops are decomposed by cooking or stomach acids, precluding their ability to cause adverse effects if ingested.

Traditional agriculture relies on crossbreeding and hybridization to improve the quality and yield of crops and domesticated animals, and to overcome natural obstacles such as disease. These methods involve controlled breeding of plants and animals with desirable traits to produce offspring that ideally will retain the best traits of the parent organisms. Virtually every plant and animal grown commercially for food or other uses is a product of crossbreeding and/or hybridization. Relative to biotechnology methods, these processes are costly, time consuming, inefficient, and subject to significant practical limitations.

Genetic modification of crops has produced herbicide resistant strains, insect resistant strains, enriched foods, and improved industrial products. Herbicide and pest resistant crops can have a profound positive impact on the environment by making it possible to raise crops using dramatically less pesticide. Transgenic corn containing insecticidal toxins from *Bacillus thuringiensis* (Bt) bacteria can prevent corn borer infestations without chemical crop dusting that is toxic to humans and also kills beneficial insects. Producing these herbicide and pest resistant crops by traditional methods—if it were possible at all—would take dozens of generations.

A study of the first ten years of commercial genetically modified crop growth—from 1996 to 2006—found economic benefits to farms of $5 billion in 2005 and $27 billion over the ten year period. Pesticide use was also reduced by 224 million kilograms, or 6.9 percent, resulting in a reduction of environmental impact by more than 15 percent.[1] While improved seeds may cost significantly more than conventional seeds, it is estimated that conventional farmers spend significantly more on chemical insecticides and herbicide than they

1 Brookes, G. and Barfoot, P. Global Impact of Biotech Crops: Socio-Economic and Environmental Effects in the First Ten Years of Commercial Use. *AgBioForum*, 2006. 9(3):139-151.

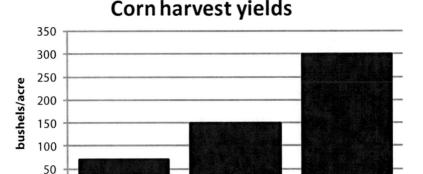

Figure 6.1 *Progress in agricultural yields*
Source: Monsanto

spend on seeds. Furthermore, genetically modified seeds have helped improve yields significantly; U.S. corn harvest yields have doubled since 1970, and Monsanto predicts that harvest yields will double again by 2030 (see Figure 6.1).[2]

The most abundant genetically modified crops are cotton, corn, soy, and canola. More than 1 billion acres of genetically modified crops had been sown in 17 countries by the end of 2005, a decade after Monsanto introduced the first genetically modified crop. According to the USDA, 52 percent of all corn, 79 percent of upland cotton, and 87 percent of soybeans planted in the United States in 2004-05 were biotechnology-derived varieties. The International Service for the Acquisition of Agri-Biotech Applications reports that in 2007 the number of farmers growing biotechnology-derived crops exceeded 12 million—11 million of whom were defined as resource-poor farmers—and hectarage exceeded 114 million acres.

CHALLENGES

Techniques to insert genes into plants are well established, but a remaining challenge for agricultural biotechnology is the realization of desired modifications. With many plants sporting genomes larger than humans, a fundamental understanding of agricultural biology is necessary for application. Confounding efforts at genetic modifi-

2 Hindo, B. Monsanto: Winning the Ground War. *BusinessWeek*, December 17, 2007.

cation, introduced genes may be unstable, and unforeseen biological issues may interfere with introduced proteins.

Another challenge for agricultural biotechnology is public acceptance. Because public support is essential to enable application of biotechnology, it is important to be sensitive to potential objections and to encourage positive perceptions. Public concerns can impact political regulations, resulting in bans on crop plantings or even sales of crops. Figure 13.1 in Chapter 13 shows the dramatic effect of an EU-wide ban on genetically modified crops on U.S. corn exports.

Critics of genetically modified foods warn that inserting genes into plants and animals may have unforeseen results, increasing the risks of allergic reactions to foods and other health problems. The potential for genetically modified plants to crossbreed with wild stocks or to cause environmental damage is also a prevalent concern. It is worth noting that many of these same concerns would also preclude the use of traditional farming practices and breeding methods if they were applied beyond biotechnology.

Beyond answering critics of genetic modification, developers must also find markets with a preference for their products and who are also willing and able to pay a profitable price. A case example of failure to find a profit-enabling market is the Flavr Savr tomato, a tomato engineered to have a longer shelf-life. In 1994 the FDA approved the Flavr Savr tomato, the first genetically modified whole food product. Interestingly Flavr Savr tomatoes were pulled from the market not because of consumer resistance, but rather due to customer disinterest and an inability to sell for a profit. See Box *Flavr Savr tomatoes: Operating in unfamiliar markets* in Chapter 11.

While improved yields, nutritional enhancement, and the ability to grow crops on marginal soils may appear to individuals in developed nations to be trivial or cosmetic improvements, but the situation in developing countries makes these improvements far more imperative. As population encroachment and environmental change are decreasing the quantities of arable land in many parts of the world, aging distribution infrastructures are challenged to deliver food to ever-increasing populations. For many developing nations the ability of biotechnology to increase crop yield, reduce farming inputs, and expand arable land presents a vital solution to the need to grow more crops on less land in order to prevent otherwise-inevitable widespread starvation.

TREE BIOTECHNOLOGY

An early goal of forest scientists has been to produce trees with less lignin, avoiding expensive and environmentally toxic procedures in paper production. Removal of lignin in paper production requires an enormous amount of energy and chemicals, making the pulp and paper industry the second most energy-intensive industry group in the U.S. manufacturing sector (see *Paper Production* later in this chapter). Another problem is that forestry demand for wood products exceeds the rate of renewal. Some tree species require well over a century to reach economic viability. Using genetic engineering to develop faster, straighter, and taller-growing trees can potentially fill a market need while preserving old growth forests.

BIOCLIP

Biotechnology also has applications in agriculture beyond growing plants and trees. Bioclip, a process that uses a naturally occurring protein that causes sheep to shed their fleece, was developed by CSIRO Animal Production in Australia and is currently being marketed. The principle is that sheep are fitted with a retaining net that assists in fleece harvesting and are administered epidermal growth factor, a protein that causes a natural break in wool fibers. Bioclip has the potential to reduce stress, cuts, and injury to sheep, while producing higher quality fleece.

ANIMAL BREEDING AND CLONING

Biotechnology can serve two roles in animal breeding. First, it can enable the use of genetic markers to identify desired animals for breeding programs. Biomarkers associated with characteristics such as milk production, meat quality, or hereditary diseases can be used to inform traditional breeding programs and improve herd quality. The second, more contentious application, is to directly clone desired animals.

Box

Carnivorous fish as vegetarians

Fish farming, or aquaculture, has been heralded as a potential solution to overfishing of wild stocks. The situation is complicated, however, by the need to satisfy carnivorous fish diets. Carnivorous fish aquaculture accounts for the majority of global fish oil usage, and is rapidly growing to become the primary market for fish meal as well—aquaculture is depleting wild fish stocks.

Enter biotechnology. By studying the protein requirements of carnivorous fish, researchers can potentially use traditional breeding and genetic modification to produce terrestrial crops as feed. These crops can both address the nutritional needs of fish, and can resolve some of the environmental issues related to aquaculture. Furthermore, growing plant crops to feed fish farms creates new revenue opportunities for farmers and holds the potential to enable inland farmers to farm fish, improving the distribution and availability of fresh fish.

Despite its promise, it remains to be seen whether traditional breeding can produce plants that can satisfy the nutritious needs of fish, if consumers will accept fish products that have been fed genetically modified plants, if the quality and flavor of plant-fed fish will match fish-fed or wild fish, and if plant-fed aquaculture can be economically feasible.

In 2008 the FDA released a "final risk assessment," concluding that foods from healthy cloned animals are as safe as those from traditionally-bred animals. Cloning animals costs substantially more than traditional or assisted breeding programs, but this additional cost is somewhat offset by the near-certainty that the clone will inherit desired traits. The extra cost of producing cloned animals also provides an assurance to wary consumers—the high cost of producing these animals means that they are unlikely to be used as meat or milk sources, but will likely be relegated to breeding.

FUNCTIONAL FOODS

Functional foods contain elements which can provide health benefits beyond simple nutrition. Existing examples include high omega-3 eggs, produced by feeding ground flax seeds to hens, and supplementing processed foods with soy protein or fiber.

Beyond improving yields, biotechnology also has the potential to dramatically impact the nutritional qualities of food. Nutritional modifications to plants include conferring the ability to synthesize essential vitamins, reducing the undesirable saturated fat content of cooking oils, increasing protein quantity and quality in vegetable staples, and reducing allergenic properties of milk, wheat, and other products.

These nutritional improvements can have a significant effect on human health. For example, enabling staple crops such as rice to synthesize vitamin A precursors or to make iron bio-available respectively hold the potential to prevent blindness and anemia in countries where commercial vitamin supplements are unaffordable. Calgene's (now owned by Monsanto) Laurical is the first commercially available functional food oil, approved by the FDA in 1995. While conventional canola oil does not contain lauric acid, Laurical contains 38 percent lauric acid. Laurical has applications in soaps and detergents, chocolate, low-fat coffee whiteners, and imitation cheeses.

MOLECULAR FARMING

Molecular farming produces useful products from domesticated plants and animals through genetic engineering. Pharming is a subset of molecular farming and produces therapeutic drugs using genetically altered animals and plants. The distinction between molecular farming and traditional farming is that the plant and animal products of molecular farming are not eaten as food, but are harvested to produce useful biotechnology products. Safeguards to prevent exposure through accidental ingestion include sequestration of crops (plans have included growing plants in abandoned mine shafts; also see Box *Starlink corn: Controlling biotechnology crops* in Chapter 8), production of recombinant proteins in non-edible portions of plants, and expressing proteins at very low levels, requiring extensive processing to obtain measurable and useful quantities of recombinant materials.

A non-therapeutic application of molecular farming is the mass-production of spider silk. Stronger than steel and lighter than cotton, spider silk manufacturing has traditionally been impeded by the inability to domesticate spiders. Nexia Biotechnologies has produced dragline spider silk in laboratory conditions. Their intention is to spin spider silk produced in the mammary glands of genetically modified goats for use in fishing line and in military applications such as lightweight body armor.

Pharming

Many developing countries that could benefit from commercially available therapies for diseases such as hepatitis and other endemic conditions cannot afford to purchase appropriate medicines or produce them locally. Production of therapeutic vaccines in familiar crops such as bananas and potatoes, or chicken eggs, can enable local farmers to manufacture medicines without the need for sophisticated production techniques or expensive purification methods.

The traditional method for manufacturing biological drugs is fermentation in huge stainless steel vats. A significant advantage of pharming is that it can decrease the cost of drug manufacturing. Furthermore, whereas scaling fermentation systems requires building and receiving FDA approval for additional facilities, boosting pharmed drug production may be as simple as planting more transgenic plants or increasing the size of a transgenic animal herd. This is especially important in countries where expertise and facilities for large-scale fermentation are not available. These advantages are offset by the higher up-front costs and longer lead-times required to produce transgenic animal or plant production systems (see *Manufacturing* in Chapter 5).

The development of drugs that are easy to purify or that can be administered without purification is essential for enabling pharming. Application of pharming is challenged by the threat of uncontrolled spread of genetically engineered plants and animals, and by start-up development costs. In the case of drug manufacturing, it is preferable to retain formulation flexibility and delay the bulk of manufacturing expenditures until a drug's safety and efficacy have been assessed. The financial cost and time required to develop transgenic production systems are at odds with this strategy.

In 2006 the European Medicines Agency (EMEA), the European counterpart to the FDA, approved GTC Therapeutics' ATryn, the first drug produced in an animal bioreactor. ATryn, a recombinant form of human antithrombin, was approved for a rare disease, hereditary antithrombin deficiency. It remains to be seen if GTC Therapeutics can expand the indications for the drug and leverage this initial proof-of-principle to produce more valuable drugs using their animal bioreactor technology platform.

FOOD SAFETY

Biotechnology can play an important role in food safety. The same technologies that detect and eliminate disease agents in the body can also be used during food processing.

Advanced molecular diagnostic techniques can detect trace amounts of impurities or infectious agents at steps throughout the food chain, allowing producers to quickly identify unsafe food and isolate the cause of contamination.

Biotechnology can also help develop new ways to keep foods safe. In 2006 the FDA approved the first use of bacteriophage as a food additive (to be sprayed on food as a disinfectant, not for nutritional purposes). Bacteriophage are a naturally-occuring type of virus that only infect bacteria. The product in this case is a mixture of six naturally-occurring bacteriophage isolates from wastewater sources, targeted at *Listeria* strains found in processed meats, and is produced by Intralytix. Food labels must include mention of bacteriophage content or treatment. Some countries have also approved bacteriophage to treat infections in humans; the FDA has not. Bacteriophage are also approved in the United States in pesticide applications.

WHITE BIOTECHNOLOGY: INDUSTRIAL PROCESSES AND BIO-BASED PRODUCTS

Industrial biotechnology is the application of molecular biology techniques to improve efficiency and reduce the environmental impacts of industrial processes. Just as biotechnology has transformed agriculture, drug discovery, and development, it can similarly affect industrial operations.

Industrial biotechnology companies develop biocatalysts such as enzymes that are used for chemical synthesis. Enzymes are a category of proteins which are produced by all living organisms (see *Proteins and Enzymes* in Chapter 3). Enzymes enable the biochemical reactions necessary for life by increasing reaction rates. In biological systems, enzymes help digest food, assemble complex molecules, and perform other complex functions. Specialized enzymes are also used extensively as detergents as well as in the production of beer, cheese, and fruit juice. Bacteria have developed specialized enzymes that allow them to live in a wide variety of extreme environments; from thermal vents at the bottoms of oceans to the insides of rocks. En-

---

Box

Blue jeans and biotechnology

1.8 billion pairs of denim jeans are sold each year, making them among the most prevalent clothing items sold worldwide. *Stonewashing* is commonly used to soften the jeans and fade the dyes to give the jeans a slightly worn appearance. This process was traditionally performed by tumbling jeans in large machines with abrasive pumice stones. This process can weaken jeans and damage washing machines, and requires several rinsings to remove all the pumice traces.

An alternative enzyme-based method has been introduced which imparts several benefits. The degree of stonewashing can be attenuated by using cellulase enzymes to break down the denim cellulose fibers in a controlled manner. This process also requires less water and energy than traditional stonewashing, and results in longer-lasting jeans.

zymes are characterized according to the compounds they act upon. Some of the most common enzymes with industrial applications are proteases, which break down protein; cellulases, which break down cellulose; lipases, which act on fatty acids and oils; and amylases, which break starch down into simple sugars.

By studying diverse bacteria and other organisms, scientists discover novel biocatalysts that function optimally under a wide variety of conditions, including the relatively extreme levels of acidity, salinity, temperature, or pressure found in some industrial manufacturing processes. In other cases, enzymes can remove the need for extreme conditions or harsh chemicals, saving energy and reducing environmental impact.

The application of biotechnology to industrial processes is appealing because of the potential to affect yield, effectiveness, and production cost of products with established markets. Serving established markets cam improve the accuracy of market size projections, helping justify high R&D investments and attracting funding for large market opportunities. The potential for application of biotechnology to reduce infrastructure requirements may also make it possible to profitably address smaller markets. For example, see the example of *Oil Well Completion* below. An additional appeal for industrial biotechnology development is the greatly reduced regulatory burden relative to pharmaceutical applications.

BIOFUEL AND LUBRICANTS

Petroleum prices, political considerations, and the threat of shortages all motivate the search for alternative fuel sources. Processes to convert cornstarch into ethanol, a petroleum additive and alternative, have been available for many years, but questions about the scalability and ultimate economics of this approach have led to the search for alternative methods to produce ethanol.

Fuels derived from petroleum are the product of compression and heating of prehistoric plants and animals over geological time scales deep below the earth's surface. Because of their ultimate biological source, the potential exists to use alternative processes to make fuels from sources such as plant materials and animal fats in

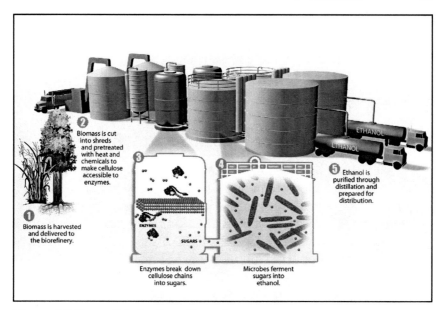

Figure 6.2 *How cellulosic ethanol is made*
Source: Genome Management Information System, Oak Ridge National Laboratory

less time. There are three basic methods to produce fuels from plant and animal sources: chemical transesterification, fermentation, and cellulose degradation. These fuels can be used in cars and trucks, as well as in numerous other applications.

Transesterification is a process used to convert vegetable oils, animal fats, and recycled greases into biodiesel. This is technically a chemical process, not an application of biotechnology. Fermentation is the use of bacteria or yeast to convert simple sugars, abundant in plants such as corn and sugar cane, into ethanol. This is fundamentally the same process that is used to make beer, wine, and other alcohols. Cellulose degradation significantly expands the prospects for fermentation by enabling the use of a wide variety of feedstocks. Whereas fermentation requires feedstocks with a high content of simple sugars, cellulose degradation uses chemical pre-treatments and cellulase enzymes to break down cellulose, a complex sugar, into simple sugars. These simple sugars can then be used in traditional fermentation to make desired chemicals.

The principal advantage of cellulose degradation over other methods is the abundance of cellulose in materials such as farm waste, wood chips, and even garbage. A majority of the material in plants is cellulose. Cellulose degradation occurs in nature, but slowly. The challenge is to increase the efficiency of cellulase, an enzyme that breaks down cellulose, and to improve the yield of cellulose

---

Box
Biofuel and bio-products: Back to the future

While the carbohydrate, or biobased, economy is sometimes touted as a modern phenomenon, it actually has a long history. In the 1800s, industrialized societies were largely carbohydrate economies. The biological basis for manufacturing proliferated well into the 20th century, but reduced access to raw materials (e.g., Japanese control of Asian rubber plantations in World War II) and inexpensive oil prices in the mid 1900s led to a shift to petroleum, or hydrocarbon, based raw materials in manufacturing and energy production. As oil prices, political considerations, and carbon dioxide release motivate the search for alternatives to petroleum, advances in agriculture, chemistry, and biotechnology have set the stage for a resurgence of carbohydrate-based manufacturing and energy production.

Carbohydrates and hydrocarbons are very similar compounds. Carbohydrates are composed of the elements carbon, hydrogen, and oxygen. Hydrocarbons are elementally simpler than carbohydrates, containing only carbon and hydrogen. Carbohydrates can readily substitute for hydrocarbons through chemical and biological manipulations which reduce carbohydrates to hydrocarbons. The challenge in such substitutions is to ensure that the manipulations are cost- and resource-efficient.

Two significant differences between the use of carbohydrates and hydrocarbons are carbon dioxide release and sustainability. Burning petroleum-based fuels releases carbon which was sequestered over thousands of years, resulting in a near-term increase in atmospheric carbon dioxide levels and depletion of slowly-replenished resources. Burning plant-based fuels releases carbon which was sequestered over the life of the plants (as short as one growing season), resulting in a negligible change in atmospheric carbon dioxide levels and consumption of a resource which can be replaced rapidly.

from biological sources.

Subsidies on corn production and tax exemptions for non-petro-leum fuels have been instrumental in enabling entrants to produce and market biofuels. The situation for biofuels is analogous to the early years of penicillin production. The basic scientific principles are known, a strong market need exists, but a better method is need-ed to enable cost-effective large-scale production.

PLASTIC

The world's first modern biorefinery, a Cargill-Dow project, went online in Blair, Nebraska in 2002. The plant is the product of a joint venture established in 1997 between Cargill and the Dow Chemi-cal Company to commercialize polylactic acid (PLA) under the brand name NatureWorks. Dow pulled out of the venture in 2004 acknowledging significant long-term potential, but dissatisfaction with short-term profitability.

PLA is made by fermenting the sugar in corn (other high-sugar feedstocks are also amenable to this process) into lactic acid mol-ecules, which are then linked to form polylactic acid. PLA can be used to make a wide array of products, including plastic cups and containers, wrappers, and polyester textiles. Furthermore, PLA is biodegradable, requires 65 percent less energy to produce than con-ventional plastics, and can reduce fossil fuel use in plastic manufac-ture by up to 80 percent.

Bio-based plastics have the added benefit of being naturally bio-degradable, reducing the environmental impact of their use. They face resistance due to their higher costs, the need to re-engineer downstream manufacturing and utilization processes in some cases, and reduced suitability in harsh environments. Despite these hur-dles, they are finding strong adoption in consumer-facing applica-tions such as disposable packaging for food containers.

OTHER BIO-BASED PRODUCTS

As described in the section *Biofuels* above, the original source of petroleum products is actually biological. Accordingly, the potential exists for biotechnology innovations to replace petroleum products

Box

Using bacteria to make snow

Snowmax is an ice-nucleating protein derived from naturally-occurring *Pseudomonas syringae* bacteria. It is hypothesized that the natural purpose of this ice-nucleating protein is as part of a long-distance dispersion strategy of *Pseudomonas syringae*, which is a plant pathogen. The bacteria are able to survive for long periods in aerial suspension. The ice-nucleating proteins help drop the bacteria out of aerial suspension, by way of rain or snow, enabling them to infect plants.

Specially-designed aeration guns are used to spray water mixed with Snowmax on ski slopes. Snowmax increases the number of nucleation centers in water droplets from these aeration guns, improving snow making efficiency and also enabling snow making at higher temperatures, ultimately saving ski resorts money and improving the quality of ski slopes. Snowmax is produced by growing *Pseudomonas syringae* in a fermentation vessel and extracting the protein using filtration processes. The product is irradiated prior to shipping to ensure that live bacteria, which are not harmful to humans, are not released.

in many manufacturing processes.

Vitamin B2 is used as a supplement in animal feed to keep animals healthy and fit. In 1990 BASF developed an innovative fermentation method to replace the traditional eight-step chemical process used for vitamin B2 production, using *Ashbya gossypii* fungus with a one-step fermentation. This fermentation process reduces costs by up to 40 percent and reduces environmental impact by 40 percent. The fermentation process has several other advantages over chemical synthesis. BASF has realized a 95 percent reduction in waste, reduced energy usage due to lower reaction temperatures, and a 60 percent reduction in the resources required.

Another product improved by biotechnology is propanediol. Propanediol is a clear colorless liquid with applications in deicing, as an engine coolant, in adhesives and coatings, and as an additive in cosmetics and shampoo. Traditionally produced from petroleum feedstocks at high temperatures, a joint venture of Dupont and Tate

& Lyle uses a proprietary fermentation method to produce a biologically-derived version of propanediol named Bio-PDO for industrial and consumer applications. The benefits of Bio-PDO over conventional alternatives are reduced production energy requirements, low toxicity, and biodegradability, and improved heat stability and reduced corrosion when used as anti-freeze. Bio-PDO is also replacing petroleum sources in the manufacture of Dupont's Sorono plastic.

PAPER PRODUCTION

Converting wood into paper is an energy, water, and chemical intensive process. Traditional processes to produce pulp for paper production use either a chemical or mechanical process. Chemical production involves boiling wood chips in a high-temperature, high-pressure, sulfide or sulfate solution. Mechanical pulp production results in lower quality pulp that is less suitable for bleaching, making it suitable for a narrower range of applications. Pulp from either source is bleached using elemental chlorine or chlorine dioxide to remove the lignin. The pulp is then refined and formed into paper. Pulping and bleaching account for the majority of pollutant releases associated with paper production.

Biopulping uses white rot fungi to selectively degrade lignin in wood, resulting in higher quality pulp, energy savings, and reduced wastewater toxicity. Using bleaching enzymes allows the process temperature to be lowered and reduces the need for rinsing. These changes can potentially increase the speed of paper production while significantly reducing the amount of chlorine and energy required, decreasing production costs and reducing emissions of toxic chlorine residues.

DETERGENTS

Detergent enzymes represent the broadest application of enzymes. Detergent enzymes improve household laundry, dishwashing, and industrial washing applications by improving cleaning performance, reducing washing times, reducing energy consumption by lowering wash temperatures, and even rejuvenating clothes.

The most common enzymes used are proteases and amylases,

which respectively remove stains and soils based on proteins and starches. Other enzymes with applications as detergent adjuncts include lipases to digest fat or oil based stains, peroxidases to inhibit dye transfer, and cellulases to prevent pilling on cotton clothes.

These innovations allow a reduction in the use of numerous environmentally-damaging chemicals, including phosphates and bleaches.

MINING

Microorganisms are used worldwide in mining processes to oxidize and leach metals. Other applications are as alternatives to harsh chemicals to remove metals from industrial wastewater streams. The primary bacteria employed are *Thiobacillus ferrooxidans, Leptospirillum ferrooxidans,* and thermophilic (high temperature) bacteria to leach metals such as copper and gold from sulfide minerals. Some of the advantages of bioleaching over conventional roasters, smelters, and pressure autoclaves are that construction time is shorter, no noxious gases or toxic effluents are produced, environmental permit and reporting processes are simpler, and safety is increased due to

Table 6.2 *Selected industrial enzymes*

Enzyme type	Function and utility
amylase	Decomposes simple sugars. Applications in textiles, laundry and dishwashing, biofuels, and paper production.
cellulase	Decomposes cellulose into simpler sugars. Applications in biofuel production, laundry, and paper processing.
laccanase	Delignification. Applications in paper processing, bioremediation, and syrup production.
lipase	Decomposes fats. Applications in laundry and surface cleaning, food processing, leather processing, and pharmaceuticals.
protease	Decomposes proteins. Applications in laundry, leather processing, baking.
xylanase	Degrades plant cell walls. Applications in paper production, biofuels, food production, and textiles.

processing at or near ambient temperatures and pressures.

Another method under development is the use of plants to mine sparse deposits of valuable minerals. Plants with enhanced abilities to sequester heavy metals in soil can extract sparse deposits of gold or other valuable minerals, which can then be recovered by simply harvesting and incinerating the plants.

OIL WELL COMPLETION

Microbial enhanced oil recovery (MEOR) is the use of microorganisms to retrieve recalcitrant oil from existing wells, maximizing petroleum production of an oil reservoir. MEOR employs the inoculation of selected natural bacterial strains into oil wells to decrease the viscosity of thick oil deposits and ease oil flow, or to produce gases such as carbon dioxide to propel oil out of the well.

BIOREMEDIATION

Bioremediation is the application of biotechnology for environmental reclamation. Some of the processes described above have applications in bioremediation. Relative to existing alternatives, the use of plants, microorganisms, and their by-products to sequester pollutants, or to degrade them into relatively benign compounds, can be a safer, cheaper, and faster method to clean the environment.

Unresolved questions regarding the release of genetically modified organisms into the environment motivate the search for more natural techniques. Fortunately, there are few natural materials that at least one naturally-ocurring microorganism cannot use as a nutrient. Given appropriate conditions, even synthetic compounds are subject to microbial metabolism. By searching for organisms already feeding on pollutants, either in natural environments or at polluted sites, it is possible to develop non-transgenic bioremediation systems with applications ranging from treatment of oil spills to reclamation of contaminated soil and water.

RED BIOTECHNOLOGY: MEDICAL APPLICATIONS

MONOCLONAL ANTIBODIES

Antibodies are natural components of human and other immune systems that recognize unfamiliar material such as infectious bacteria and cancerous cells and help eliminate them. While our natural complement of antibodies is generally very effective at recognizing and prompting the destruction of infectious microorganisms and cancerous cells, threats are sometimes missed. Monoclonal antibodies are a category of biotechnology-derived drugs that are designed to act and look like naturally occurring antibodies and may directly treat diseases or condition a patient's own immune system to launch a highly specific attack on infections or diseased tissues. They are designated "monoclonal" because they are produced as large batches of identical molecules. Georges Köhler and César Milstein received the 1984 Nobel Prize in Physiology or Medicine for their description of a technique to produce monoclonal antibodies. They shared the prize with Niels Jerne, who described the development and control of the immune system.

In the 1980s the first antibody trials saw early experimental therapies rendered inactive by the liver, or activating patients' own immune systems to raise antibodies against the foreign therapeutic antibodies, resulting in increased illness. The rejection of these initial non-human antibodies can be attributed to the primary purpose of the immune system: to repel foreign bodies. The use of modified versions of animal antibodies, humanized antibodies, and fully human antibodies led to the development of monoclonal antibody therapies that are safe and effective. Genentech's Rituxan was the first monoclonal antibody to be approved for cancer treatment in the United States. Rituxan works by binding to specific types of cancer cells and triggering the immune system to destroy them. Another Genentech product, Herceptin, is targeted at growth factors that are directly implicated in approximately 20 percent of breast cancers (see Box *Personalized medicine and drug sales* later in this chapter).

While the target specificity of antibodies offers significant therapeutic benefits, the size and characteristics of these massive proteins present a number of unique challenges. The large size of antibodies precludes oral dosage, the preferred method of drug administration. Long-term storage is also an issue, as the structural integrity of antibodies must be maintained. While some drugs can exert their therapeutic activity within seconds, antibodies may take up to a week for therapeutic effect because of the time needed to traverse the bloodstream, find, and then bind their molecular targets. Additionally, antibodies do have the potential to bind to more than one target, potentially leading to serious side effects.

Because all antibodies share common structural elements, the relative similarity of antibodies and of their manufacturing techniques also present benefits and challenges for drug developers. Early indications suggest that approval rates for monoclonal antibodies as a group are higher than average drug approval rates, likely due to the similarity of manufacturing techniques and safety and efficacy assays (or, conversely, the prevalent application for pressing needs such as cancer which employ short-term use and often have relaxed side effect standards). The challenge emerging from this homogeneity is that patents exist for virtually all the steps, and the alternative methods, for the production of antibodies. An antibody producer that can navigate this web of patents and also produce a viable drug must contend with the prospect of stacked royalties due to all the patent holders (see *Stacking royalties and submarine patents* in Chapter 7).

The target specificity of antibodies also poses a challenge for generic producers. Subtle differences in recognition sites or antibodies structure may have profound though difficult-to-predict consequences. Whereas generic manufacturers of small-molecule therapeutics need only prove their version biologically equivalent to pioneer versions, it is not currently possible to prove bioequivalence for an antibody (see *Generic Drugs* in Chapter 8). This raises a significant barrier to entry by competitors and potentially permits differentiation among competing antibodies.

Antibody-based drugs can enjoy years of strong sales with minimal competition, even after patent expiration, because of the difficulty of demonstrating equivalence of antibodies produced by a second party. This challenge, however, motivates competitors to produce improved antibodies rather than simply producing competing antibodies. This means that when an innovator's antibody is challenged by a new entrant, the innovator is likely to lose a greater portion of market share than they might if the competing drug were merely equivalent to the original.

RNA INTERFERENCE

RNA interference therapies, sometimes referred to as antisense therapies, block gene expression. Summarizing information presented in Chapter 3, mRNA is a molecule that transfers information from genes to the protein synthesis machinery within cells. The goal of RNA interference is to intercept an mRNA message before it is translated into protein. Andrew Fire and Craig Mello shared the 2006 Nobel Prize in Physiology or Medicine for their discovery of gene silencing by RNA interference.

It is important to recognize that RNA interference is a subtractive solution. Interference cannot directly replace, amplify, or add a gene function. It can only inhibit a gene function (although inhibiting an inhibitory gene can indirectly increase the expression of a second gene). RNA interference has the potential to treat a range of diseases including cancer, autoimmune disorders, and infectious diseases.

A successful RNA interference therapy must enable entry of interfering molecules into cells to permit a therapeutic effect, prevent degradation of interfering molecules before they can act, and ensure specificity so that essential functions are not disrupted. The relative ease of controlling these issues in laboratory settings means that many RNA interference therapies that look promising in preclinical development are likely to face complications in therapeutic settings. Isis Pharmaceuticals' Vitravene, the first RNA interference drug, overcomes delivery issues through direct injection into the target tissue.

RNA interference therapy is also limited by the molecular understanding of disease processes; knowledge of which genes to block and the side effects of inhibiting their expression are essential in selecting appropriate RNA targets. Careful dosage is also critical, as delivery of too much interfering RNA has been shown to overload

RNA processing systems in test animals, killing them.

GENE THERAPY

Gene therapy uses genes to treat disease. Techniques for gene therapy include replacement of defective genes, and supplementation with therapeutic genes. For diseases caused by an absent or defective copy of a specific gene, supplementation of that gene can potentially cure the disease. The technical challenges of gene therapy include targeting appropriate cells and tissues, ensuring gene transfer, controlling gene expression, and satisfying safety concerns.

While most genetic deficits require gene expression in specific cell types, some diseases can be cured by expression of specific genes in a variety of cells. Blood-related deficiencies, such as the lack of clotting factors in hemophilia, can potentially be cured by enabling the cells lining blood vessels to produce the necessary clotting factors. A caution for untargeted therapy is that certain cell types may suffer complications from expression of inappropriate genes. Another potential complication is that a patient's immune system may reject an introduced gene product and the cells producing it, leading to destruction of healthy tissue.

Regulation of quantity and duration of gene expression is also important. While diseases such as cystic fibrosis and hemophilia require persistent expression and may be cured even with low expression, other diseases such as diabetes require tightly regulated and coordinated gene expression. For some genetic diseases, irreparable damage occurs early in life. For example, cystic fibrosis leads to lung damage during childhood. It is important to intervene and treat such diseases before permanent damage is sustained.

An early success for gene therapy was witnessed in a 1990 trial when two girls with a genetic deficit causing severe immunodeficiency were given infusions of their own immune system cells. These cells were genetically engineered to contain a working version of their missing gene. Following regular monthly administration, the girls developed active immune systems that allowed them to remain healthy for more than 10 years.

The first marketed gene therapy product was approved in China in 2003. Shenzen SiBiono's Genicide is targeted at head and neck squamous cell carcinoma, a highly lethal cancer with an annual incidence of 300,000 people in China. The drug uses a benign viral vector to deliver *p53*, a gene implicated in controlling cell growth; many tumors contain defective *p53* or fail to express sufficient quantities of the protein.

Despite its early promise, advances in gene therapy have been hampered by variability in the safety and effectiveness of trials, sometimes with fatal consequences.

DIAGNOSTIC TESTS

In addition to treating diseases, biotechnology has also made it easier to detect and diagnose medical conditions. A quantum leap past traditional techniques that require correlation of numerous symptoms to develop a diagnosis, biotechnology enables the direct detection of biological processes. In addition to refining symptom-based diagnoses, it is also possible to make determinations at earlier stages. Screening for pregnancy and cancer are examples of diagnoses that have increased in reliability and sensitivity as a result of biotechnology.

For diseases where symptoms are usually noticed past the point where treatment is most effective, diagnostic tests can save lives by enabling at-risk individuals to monitor their health prior to onset of disease. Individuals with genetic predispositions to specific cancers can be alerted to their increased likelihood of disease and can engage in preventative activities and regular screenings, potentially avoiding disease progression or allowing early intervention.

A secondary benefit of diagnostic tests is that they can enable individuals to avoid costly, dangerous, or unnecessary procedures (see Box *Personalized medicine and drug sales* later in this chapter). For example, use of aspirin to prevent colorectal cancer may be less cost-effective than regular screening. Citing a cost of nearly $150,000 per year of life saved by preventative use of aspirin, factoring in the costs of the drug and treatment of side effects, versus a $30,000 cost of screening per year of life saved, a recent study concluded that screen-

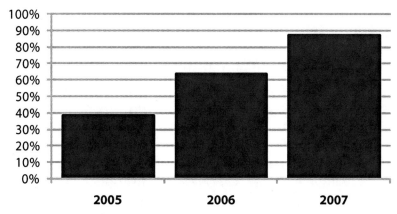

Figure 6.3 *U.S. babies born in states mandating genetic disorder testing*
Source: March of Dimes

ing was preferable to aspirin use.[3]

On initial examination, diagnostics may seem to be a good market-entry objective for start-ups seeking to develop an initial revenue stream. Diagnostics are relatively cheaper to develop than therapeutics and it is also relatively easier to gain regulatory approval for them. However, because drugs serve more pressing needs than diagnostics they can be assured of relatively greater sales and greater profits due to decreased price elasticity.

PERSONAL GENETIC PROFILING

As information about genetic markers increases and DNA profiling and sequencing costs decrease, numerous companies are entering the personal genetic profiling space. These companies offer to profile or sequence an individual's DNA and identify predispositions to disease, with the objective of enabling their customers to take proactive approaches to prevent disease onset.

While the potential to use genetic information to help prevent disease is an exciting prospect, the science behind predicting disease predispositions is too immature to be useful in most cases. A federal panel investigating personal genetic tests concluded that "a growing number of the tests are being marketed with claims that are unproved, ambiguous, false or misleading" and added that a potential

---

3 Ladabaum, U., *et al.*, Aspirin as an adjunct to screening for prevention of sporadic colorectal cancer. *Annals of Internal Medicine*, 2001. 135(9):769-781

for harm exists "if a test is inaccurate, patients may be given risky, unnecessary treatments or denied treatments that would be highly beneficial."[4]

Aside from fraudulent operations seeking to sell vitamin mixtures of dubious benefit to consumers, the very science of disease predisposition mapping is in its infancy. Many diseases have strong environmental and genetic components, and the influences of these factors need to be understood for genetic profiling to be useful. Faced with a poorly-documented association of a genetic sequence with a disease, or conflicting reports of genetic predisposition, testing companies and patients are likely to assume a stronger connection than is warranted by the data.

Several state and federal agencies, including the FDA, have the power and mandate to regulate genetic tests, but they are challenged to keep up with the increasing number of test providers, and to find ways to regulate tests without stifling innovation. The situation for personalized genetic testing today is analogous to the drug industry before the FDA mandated safety and efficacy testing. Companies with valid products must contend with a mass of unproven science and firms making unfounded claims. Improved science and a coherent regulatory framework are necessary to validate the field.

PERSONALIZED MEDICINE

Personalized medicine involves the application of technologies such as functional genomics (see Chapter 5) to tailor therapies to the patients most likely to benefit from them. It is estimated that most commonly used drugs are effective in only 30–60 percent of patients

4 Pear, R. Growth of genetic tests concerns federal panel. *New York Times*, January 18, 2008.

Box

Personalized medicine and drug sales

Personalized medicine has great potential to improve the safety and efficacy of drugs by targeting therapies to those most likely to benefit from them and excluding patients who are susceptible to deleterious side effects. While patients generally stand to benefit from personalized medicine, biotechnology companies may benefit or suffer based on whether screening methods are used to identify, or to exclude, patients. The contrasting examples of Herceptin and Aczone demonstrate how personalized medicine can benefit or hurt drug sales.

Herceptin

Genentech's Herceptin is a monoclonal antibody directed at the Her-2 cell receptor. Overexpression of Her-2 is implicated in approximately 20 percent of breast cancers. It is estimated that without a test for Her-2 overexpression, Genentech would have needed to perform clinical trials on 2,200 patients for ten years in order to demonstrate the efficacy of Herceptin.[1] Utilizing a test for overexpression of Her-2 to segregate patients, Genentech was able to demonstrate Herceptin's ability to safely increase survival times by 50 percent using only 469 patients in less than two years.

Because the test for Her-2 overexpression is tied to the mode of action of Herceptin, patients are able to avoid many of the unnecessary side effects associated with ineffective medicines and may benefit from early prescription of the drug most likely to effectively treat their tumors, while Genentech benefits from preferred prescription to patients who test positive for Her-2 overexpression.

Aczone

QLT's Aczone is a topical drug used to treat acne. In the course of clinical trials it was found that people with a blood disorder called G6PD deficiency have a higher risk of developing anemia with Aczone; roughly 1.4 percent of patients in clinical trials had this disorder. The potential for anemia among patients with G6PD deficiency taking Aczone spurred the FDA to require that patients be tested for the enzyme deficiency prior to being prescribed Aczone.

Unlike Herceptin, where the diagnostic test is optional prior to prescription and identifies the target population, Aczone prescription requires prior testing to exclude patients likely to suffer deleterious side effects. The impact of this excluding screen is that while patients are protected from adverse reactions, QLT suffers a significant barrier to prescription.

---

1 Tansey, B. Genentech a big believer in diagnostics, *San Francisco Chronicle*, May 17, 2004. p. B-1.

with a given disease. A subset of these patients may suffer severe side effects. There are two causes for this difference in response. First, most diseases have myriad causes. They tend to be defined by symptoms, but each distinct cause may respond best to a different treatment. Second, people are different. Differences in liver metabolic enzymes, for example, can determine if patients are unlikely to respond to a drug or if they will suffer severe side effects—see the section *Pharmacogenetics* in Chapter 5 for more details.

Aligning treatments with the patients most likely to benefit from them holds the potential to improve the effectiveness of medical intervention, while reducing healthcare costs and dangers. The key to realizing personalized medicine is alignment of a diagnostic test with a therapeutic intervention. A well-paired test, such as the test for Her-2 overexpression which indicates that Genentech's Herceptin is the preferred drug, can have a strong positive impact on sales. In other cases, personalized medicine may mean fewer sales.

While fewer sales may represent a barrier to implementation of personalized medicine in some cases, it can also be a benefit. Using profiling in clinical trials can speed approval. It is estimated that the time saved in Herceptin's clinical trials netted Genentech between $1.2 and $1.5 billion. Selling drugs to large populations also greatly increases the likelihood that significant side effects will emerge and may potentially lead to market withdrawal and debilitating lawsuits. The example of Vioxx (see Box *Vioxx: Anticipating and disclosing side effects* in Chapter 8) illustrates the risks encountered in serving large populations.

Two other barriers are physician practices and reimbursement. If diagnostic tests significantly complicate the prescribing patterns of physicians without clear benefit, or if they are not are mandated by payers, then they are unlikely to be used. The Box *Biotechnology myth: Build it and they will come* in Chapter 13 illustrates the challenges of selling diagnostic tests which require changes in physician practices. Reimbursement is described in greater detail in Chapter 13.

TISSUE ENGINEERING

Tissue engineering is the production of natural or synthetic organs and tissues which may be implanted as fully functional units, or as tissue which undergoes further development following implantation to perform necessary functions. The first engineered tissues were skin equivalents used to treat burn victims, and structural scaffolds to replace heart valves, arteries, and bones. Alternative treatments for tissue and organ failure include transplantation from donors, surgical repair, artificial prostheses, mechanical devices, and in a few cases, drug therapy. Tissue engineering has the potential to provide an alternative or complement to these treatments, potentially with fewer side effects and a greater ability to treat major damage.

While some cell types and tissues are amenable to production in liquid media or on solid surfaces, a challenge for production of more complex tissues and organs is the development of appropriate scaffolds to model growth and methods to direct local differentiation of tissues. Large organs must be perfused by blood vessels to allow for oxygenation, delivery of nutrients, and removal of waste. An alternative to laboratory production of implantable organs is the use of stem cells, which can potentially be coaxed to repair tissues upon injection into patients.

Tissue engineering also faces significant commercial challenges. Stem cells and xenotransplantation offer alternative methods to serve many of the same markets as tissue engineering. The history of tissue engineering also serves as an example of how quickly fortunes can turn in biotechnology. Despite research and development expenditures of $4.5 billion, as of 2002 none of the tissue engineering products on the market were profitable.[5] By 2007, the sector consisted of 50 profitable firms which had treated over a million patients and were generating $1.3 billion in annual sales.[6]

---

5 Lysaght, M.J., Hazlehurst, A.L. Tissue engineering: the end of the beginning. *Tissue Engineering*, 2004. 10(2):309-320.

6 Lysaght, M.J., Jaklenec, A., Deweerd, E. Great expectations: Private sector activity in tissue engineering, regenerative medicine, and stem cell therapeutics. *Tissue Engineering Part A*, February 1, 2008. 14(2):305-315.

STEM CELLS

Stem cells can repair damaged organs and tissues and even have the potential to produce entire organs in laboratory settings for use as human replacement parts. Understanding and controlling the ability of stem cells to repair organs holds the potential to eliminate the need for transplantation and tissue engineering. Unlike most of the cells in human adults, stem cells are able to differentiate into other cell types. Degenerative diseases such as Alzheimer's disease and Parkinson's disease, as well as diseases marked by cell injury or malfunction, such as stroke, heart attack, cancer, and spinal cord injury, are all candidates for stem cell therapy.

Stem cells can be harvested from remnants of fertility treatments, placentas, umbilical cords, and from adult tissues. Stem cells from embryonic sources such as discarded fertilized eggs from fertility treatments carry a significant ethical and political burden. Additionally, there are restrictions on the use of federal funding for embryonic stem cell research in the United States (see Box *State autonomy and federal laws* in Chapter 9). Because of the central importance of federal funding in supporting basic research (see Figure 4.3) and the potential for political sentiments to translate to regulatory decisions, the potential of embryonic stem cell research in the United States is uncertain.

Stem cells extracted from adult sources such as bone marrow do not share the same ethical and funding issues as embryonic stem cells, but it is clear that they have different properties as well. There are three categories of stem cells, distinguished by their ability to differentiate into other cell types. Totipotent cells can grow into an entire organism. Embryonic stem cells are unique in having this property. Pluripotent cells cannot grow into a whole organism, but they are able to differentiate into the various cell types of the body and potentially form organs. Multipotent (sometimes called unipotent) cells can only form certain types of cells such as blood cells or bone cells. Embryonic stem cells are pluripotent and multipotent by virtue of their totipentcy, while adult stem cells may be pluripotent or multipotent depending on their source.

A market for leukemia treatment with adult stem cells already

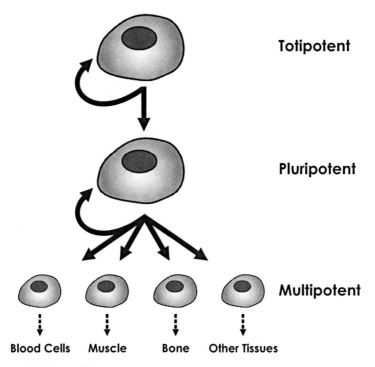

Totipotent

Pluripotent

Multipotent

Blood Cells   Muscle   Bone   Other Tissues

Figure 6.4 *Stem cell types*
Modified from National Institutes of Health

exists. A growing body of research also indicates that adult stem cells can cure hearts and other organs that have suffered trauma. Interestingly, implantation of embryonic stem cells has been implicated in cardiac arrhythmias and cancer development, suggesting that adult stem cells may be better suited for these applications.

Stem cells are also being explored as a vehicle to deliver genes to specific tissues in the body for gene therapy or cancer treatment. The greatest technical challenge facing stem cell researchers is elucidating the factors that activate stem cells to form specific kinds of tissue. Isolating and purifying sufficient quantities of stem cells for clinical use presents an additional challenge. Extensive patenting also makes freedom to operate an important consideration.

XENOTRANSPLANTATION

Shortages of human organs available for transplant and disease considerations have prompted the search for alternative sources of organs. Xenotransplantation is the transplantation of organs from any other species into humans. The similarity of human and pig organs has led most xenotransplantation research to be focused on pigs. Immune rejection and the potential for novel infections to spread into human populations are the most pressing challenges facing xenotransplantation.

Xenotransplantation need not involve implantation of whole organs. Studies have shown survival of fetal pig neural cells when administered to patients with Huntington's or Parkinson's disease. The potential also exists to coax such cells to perform needed functions. It is also possible to use animal organs for *ex vivo* treatments; use without implantation. In 1997, a patient with liver failure had his blood perfused through a liver from a transgenic pig raised by Nextran (Baxter donated Nextran to the Mayo Clinic in 2003), keeping him alive for over six hours until a liver donor could be found.[7]

Rejection may occur by multiple mechanisms with fast or slow time profiles. Methods for dealing with rejection range from administering immunosuppressive agents to removing immunoreactive elements through genetic engineering of donor animals. Aside from rejection, a significant challenge for xenotransplantation is the prospect of novel infections being introduced through the transplant recipient into the human population. One way to partially alleviate this risk is to genetically engineer animals to remove known risk factors.

It is important to note that there are alternatives to xenotransplantation that avoid immune system rejection and disease issues. Tissue engineering and stem cells can potentially permit the growth of organs and tissues outside or inside the human body, eliminating the need for xenotransplantation.

---

7 Stolberg, S.G. Could this pig save your life? *New York Times,* October 3, 1999.

III

Laws, Regulations, and Politics

Product development in biotechnology relies on innovative research and development that expands the realm of scientific knowledge. The innovative nature of biotechnology research and the potential to do harm place significant burdens on biotechnology development. Unique legal, regulatory, and political issues differentiate biotechnology from other businesses. For example, new product development is expensive and time consuming whereas in many cases copying existing products is relatively straightforward. In order to support innovation and enable companies to recoup their research and development investments, intellectual property protection is required to maintain a competitive advantage.

Because many biotechnology products are based on existing life forms and may be intended for medical use by humans or exposed to the environment, it is very important to ensure that applications of biotechnology will not do harm. Government agencies such as the Environmental Protection Agency, Department of Agriculture, and Food and Drug Administration oversee biotechnology product development. Genetically modified microorganisms, plants, and animals must be proven to be environmentally safe before they can have an opportunity to interact with the environment. Drugs meant for human use must be thoroughly tested for safety and efficacy in humans before they can be granted marketing clearance.

Because biotechnology involves manipulations of living things, political and ethical decisions have an important impact on the ability to perform research and commercialize certain inventions. Bans on technologies such as cloning, stem cell research funding restrictions, and boycotts of genetically modified food can eliminate markets and influence the scope of scientific research.

Chapter 7

Intellectual Property

> Trying to patent a human gene is like trying to patent a tree.
> You can patent a table that you build from a tree, but you
> cannot patent the tree itself.
> *William Haseltine, former Chairman and CEO, Human Genome*
> *Sciences*

Intellectual property protection is essential in biotechnology because while the cost of innovation is high, the cost of imitation is relatively low. Unlike commodity-based industries, where access to cheap materials, labor, or markets can provide a competitive advantage, innovation-based industries such as biotechnology rely on the ability to generate and exploit knowledge to gain a competitive advantage. Research and development must be financed by sales, which can only occur after research and development have been completed. Intellectual property protection is necessary to secure a competitive advantage and ultimately promote innovation, enabling innovators to prevent competitors from offering prices that reflect their reduced R&D burden, Intellectual property protection therefore plays an integral role in enabling biotechnology research by establishing a barrier to competition that permits pioneers to sustain lengthy research efforts and recoup their research and development

Table 7.1 *Intellectual property rights*

Patent	Prevent others from practicing an invention
Trade secret	Protect information and know-how
Trademark	Protect company and product name, look, and feel
Copyright	Protect the products of ideas

103

costs.

Intellectual property differs from other forms of property because it is the product of intellectual effort and may be embodied in a concept rather than a physical representation. There are four types of intellectual property rights: patent, trade secret, trademark, and copyright. Patents allow an inventor to prevent others from practicing an invention without permission. Trade secrets are unique among intellectual property rights in that they protect information and know-how that are not in the public domain. Trademarks are words, symbols, or phrases used to identify a particular manufacturer or seller and their products and to distinguish them from others. Copyrights protect the products of ideas—books, art, movies, etc.—but not the ideas themselves.

PATENTS

U.S. patent law grants the right to exclude others from making, using, offering for sale, or selling an invention in the U.S. Patent laws vary in different countries. Independent patent applications must be filed for each country in which protection is desired, although regional patent offices such as the European Patent Office enable patent protection in multiple countries. The commentary in this section is specific to the U.S. patent system, with important differences in other countries highlighted as necessary.

It is important to understand that patents do not grant the right to practice an invention, only the ability to exclude others from doing so. Concerns over patent holders "owning" genes arise from a misunderstanding; patents only grant the right to exclude others from novel applications of genes, patents do not assign ownership or the right to use a gene. Furthermore, patents cannot be used to protect naturally-occurring processes.

The objective of patent grants is to provide an incentive for innovation, and to reward inventors for publishing the best method to practice an invention. In return for disclosing the best mode for practicing an invention, inventors are given protection from competition for a number of years. In industries such as biotechnology where reverse-engineering if often relatively simple, patents are essential to

Table 7.2 *Top biotechnology patent holders*

Company	Number of patents (1976-2005)	2007 revenues ($millions)
Monsanto	3,763	8,563
Genentech	983	11,724
Chiron	834	1,920[1]
Amgen	699	14,771
Isis	698	70
Incyte	567	34
Human Genome Sciences	458	42
Millenium	424	528
ZymoGenetics	345	38
Genencor	320	410[2]

1. 2005 revenues; Chiron was acquired by Novartis in 2006.
2. 2004 revenues; Genencor was acquired by Danisco in 2005.

Source: Aggarwal, S., Gupta, V., Bagchi-Sen, S. Insights into US public biotech sector using patenting trends. *Nature Biotechnology, 2006.* 24(6):643-651.

motivate innovation. Without patent protection, innovation of products subject to reverse-engineering would be discouraged by the inability to recoup research and development investments.

There are two categories of patents that are applicable to genetic engineering and biotechnology: utility patents and plant patents. A third category, design patents, protects ornamental designs. Patents filed prior to June 8th 1995 last the longer of 17 years from the date of issue or 20 years from the date of filing. The term of plant and utility patents filed after June 8th 1995 is 20 years from the date of filing.

In the U.S. it is not permissible to patent something that was patented or described in a printed publication in any country more than one year before the date of filing in the U.S., or which was in public use or sale in the U.S. more than one year prior to filing the patent application. Because this one-year grace period is not available in many other countries, many American companies delay any publication, public use, or demonstration of inventions until after filing for patents, thereby enabling international patent application filings.

When a utility or plant patent expires, the subject matter of the patent becomes part of the public domain. This is one of the sacrific-es that are made in exchange for patent protection. An alternative to patent protection is to maintain an invention as a trade secret. While trade secrets do not have expiration dates, they offer no protection from competitors who may reverse-engineer or independently de-velop an invention; the potential to retain exclusive use of an inven-tion for a longer term than granted by patent protection is countered by the potential for reverse-engineering and independent invention. The decision to patent an invention or maintain it as a trade secret requires consideration of the ease of reverse-engineering and inde-pendent discovery versus the benefits associated with the right to exclude others from use of the invention.

PATENTING BIOTECHNOLOGY

Typical of innovative industries, a flood of patent applications has been filed in an attempt to secure of biotechnology inventions. Patent offices worldwide are challenged to interpret their patent guidelines to determine the validity of biotechnology patent claims. A case that arguably permitted the development of the biotechnol-ogy industry in the United States was settled in 1980 when the Su-preme Court reaffirmed in *Diamond v. Chakrabarty* that genetically modified organisms were patentable.

Working as a General Electric research scientist, Ananda Chakrabarty combined the individual genes that conferred the abil-ity to break down discrete components of crude oil, all within a single bacterial strain. The original patent application consisted of 36 claims to genetically engineered bacteria containing elements en-abling them to degrade crude oil, permitting application to cleaning up oil spills. The examiner of Chakrabarty's patent allowed claims detailing the production and preparation of the bacteria, but did not permit claims to the bacteria themselves, claiming "as living things, microbes are not patentable subject matter." The Supreme Court dis-agreed, referring to the Patent Act of 1952 in which Congress intend-ed statutory subject matter to include "anything under the sun that is made by man." Accordingly, Chakrabarty's oil-eating bacteria, as the product of human manipulations, were deemed eligible for patent protection.

The Supreme Court's decision allowed for the protection of re-combinant organisms and the products they produce. In 1987 the Patent and Trademark Office (PTO) further defined the scope of patents by announcing that it considered "non-naturally occurring, non-human, multi-cellular living organisms, including animals, to be patentable subject matter."

What is Patentable?

Biotechnology inventions may claim a novel product—a new "manufacture" or a new "composition of matter"—as well as applications of the resulting materials in therapeutics or diagnostics. Raw products of nature are not patentable. They are not novel. However, a new use for a known product may be claimed in a patent. It is also possible to patent novel and useful processes, or improvements on existing processes. Process claims, like product claims, cannot refer to naturally occurring instances. Furthermore, a synthetic composition is not patentable if it is identical to naturally occurring products.

As stated in 35 U.S.C., patentable biotechnology inventions include:

> A process of genetically altering or otherwise induc-ing a single or multi-celled organism to:
> - Express an exogenous nucleotide sequence
> - Inhibit, eliminate, augment, or alter expression of an endogenous nucleotide sequence
> - Express a specific physiological characteristic not normally associated with that organism
>
> Cell fusion procedures yielding a cell line that ex-presses a specific protein (e.g., monoclonal antibody)
> A method of using a product produced by the above manipulations

Therefore, while the discovery of a newly identified product of nature is not patentable, the extraction or isolation of substances from their natural environment, to make them available in a useful form for the first time, is patentable. This allowance is the basis of many "gene" or "naturally-occurring" drug patents. The gene or naturally occurring substance itself is not the subject of the patent; the processes for purification and alterations are the protected elements. Patents on genes generally claim either recombinant or purified and isolated forms of a gene, or applications of a gene. The same holds for drugs which are based on biological extracts. One cannot patent such a drug, but it may be possible to patent the purification, delivery method, or novel applications of the drug.

Patents must also define a specific application. It is not sufficient to say that a patent can be used in a broad array of situations. Applicants are required to demonstrate the diversity of applications. In the 1999 case of *Enzo Biochem v. Calgene Inc.*, Enzo's patent asserted that their antisense technology was "broadly applicable with respect to any organism containing genetic material." The court ruled against Enzo after determining that they did not sufficiently provide direction or examples of how to practice the patented claims in diverse cell types.

UTILITY PATENTS

Utility patents protect new and useful processes, machines, products, compositions of matter, or any new and useful improvement thereof. There are three basic requirements for utility patents: nonobviousness, novelty, and utility.

The criteria to satisfy non-obviousness and novelty requirements are dynamic, reflecting the expansion of knowledge in pertinent fields and further complicating the establishment and defense of patents. Patent laws are also subject to change, expanding or reducing the scope or terms of patents. The requirement for patent utility claims to be substantial, in addition to the requirements of specificity and credibility, was recently added to ensure that claims are in context with the nature of an invention.

Non-obviousness is judged by taking the frame of mind of an average person in a given field, with knowledge of all prior art. Prior art is the public knowledge that exists in a field. The existence of any prior art demonstrating that an invention is either not new or is obvious makes the invention ineligible for patent protection. Generally, if something provides new and unexpected results, it may be patentable. Novelty simply implies that something which is already known or patented cannot be patented.

In order to be deemed useful, a patent must describe a substantial, specific, and credible application. Utility patents do not apply to theoretical phenomena or ideas.

SUBSTANTIAL UTILITY

On January 5th 2001, the PTO issued guidelines specifying that patentable inventions must be substantial in addition to the previous requirements of specificity and credibility. The requirement for claims to be substantial means that it is not possible to patent something that is potentially useful but has not been fully investigated. The utility must define a real-world context of use and must be consistent with the properties of an invention. One cannot patent a potential anti-cancer drug for use as landfill or a genetically engineered mouse as snake food in order to avoid demonstrating utility.

ENABLEMENT

The purpose of patents is to give inventors a temporary monopoly in exchange for developing new innovations and for disclosing the methods by which to practice those inventions. Patents must describe an invention in sufficient detail to enabled an appropriately skilled individual to create and use the invention. Lack of enablement can limit a patent's strength. Improperly described processes or claims cannot be protected. Several examples of patent challenges which failed due to lack of enablement are described in the *Reach-Through Claims* section later in this chapter.

The requirement for inventors to describe the best mode to practice an invention is unique to the United States, and prevents inventors from gaining a patent without fully returning the societal benefit of disclosing the best method to practice the protected invention. While making production methods public may enable competitors to adapt them for current use or to develop competing products when a patent expires, incomplete disclosure can invalidate a patent.

Box
ESTs lack utility and enablement without proven use

In the case *In Re Dane K. Fisher and Raghunath V. Lalgudi*, the U.S. Court of Appeals for the Federal Circuit (CAFC) considered whether a patent application for expressed sequence tags (ESTs), which are alleged to be useful as research tools, satisfied the utility and enablement requirements. Specifically, the U.S. PTO rules require a patent application to disclose a specific, substantial and credible utility and to be enabled, which requires an application to show others how to make and use an invention without undue experimentation.

Monsanto had appealed the rejection of its patent application for five ESTs that encode parts of genes whose functions are unknown. The U.S. Patent and Trademark Office refused to grant the patent on the grounds that the ESTs had no specific utility. While the application disclosed that the ESTs may be used in a variety of ways, including use for the identification of polymorphisms and use as probes or as a source for primers, there were no supporting experimental data for any of these uses. The examiner concluded that the disclosed uses were not specific to the claimed ESTs, but instead were generally applicable to any EST.

In affirming, the CAFC agreed that each of the claimed ESTs lack a specific and substantial utility and that they are not enabled, in part because the function of the underlying protein-encoding gene was not identified. Essentially, the court felt that the claimed ESTs act only as research intermediates that may help scientists to isolate the particular underlying protein-encoding genes and conduct further experimentation on those genes.

An asserted use must not be so vague as to be of no value. The court felt that nothing about the alleged uses set the claimed ESTs apart from the more than 32,000 ESTs disclosed in the application or from any EST derived from any organism and that the application only disclosed general uses for the claimed ESTs, not specific ones that satisfy 35 U.S.C. § 101.

ESTs are, in the words of the Supreme Court, mere "object[s] of use-testing" upon which scientific research could be performed with no assurance that anything useful will be discovered in the end. Accordingly, any patent application claiming an EST should include some data identifying at least one credible use of the EST.

Contributed by Stephen Albainy-Jenei of Frost Brown Todd LLC: www.patentbaristas.com

Even trade secrets must be disclosed if they are part of the best mode of practicing an invention.

Invalidation based on failure to publish the best mode entails a two-pronged test. First, it must be determined if, at the time of invention, the inventor knew of a method of practicing the invention which he/she felt to be superior to the published method. The second test asks whether or not the patent adequately described the best mode to practice the invention.[1]

Non-obviousness

In the absence of prior art of other clear demonstrations of patent invalidity, challengers may assert that a patent was obvious at the time of invention. This is not a trivial challenge—the challenger must present compelling evidence of the obviousness of the invention. A common method to rebuff obviousness challenges is to provide examples of previous failed attempts to solve the problem addressed by the patented invention. Furthermore, if an invention is a commercial success it can be argued that other parties had strong financial motivations to address the problem solved by the patent but were unable to do so. Unexpected benefits, such as a drug combination with greater-than-expected effectiveness, novel properties, or a less-than-additive side effect profile, can also be used to defend the non-obviousness of an invention.

The bar for demonstrating non-obviousness was raised by the April 30th 2007 ruling in *KSR Int'l Co. v. Teleflex Inc*, where KSR was challenging a Teleflex patent combining adjustable gas pedals and electronically actuated fuel systems in automobiles. This case redefined the standard of obviousness by gauging the predictability of an invention. The Supreme Court argued "if a technique has been used to improve one device, and a person of ordinary skill in the art would recognize that it would improve similar devices in the same way, using the technique is obvious unless its actual application is beyond that person's skill" and concluded that "granting patent protection to advances that would occur in the ordinary course without real innovation retards progress and may ... deprive prior inven-

---

1 Tsao, Y.R., Tabtiang, R.K. Putting your best mode forward. *Nature Biotechnology*, 2000. 18:1113-1114.

tions of their value."

The previous requirements to successfully challenge the obviousness of a patent were to show that all parts of a claimed invention were known previously and that there was prior "teaching, suggestion, or motivation" to combine these prior technologies to produce the invention. The *KSR Int'l Co. v. Teleflex Inc.* ruling modified the requirement to show teaching, suggestion, or motivation with evidence of the predictability of an invention. This reduced burden to challenge a patent makes it easier to successfully challenge patents, and also places an increased burden on innovators to demonstrate the non-obviousness of their inventions.

PLANT PATENTS

Plants are granted special patent status. The original patent act of 1790 provided no protection for plants or animals, yet it was possible to generate valuable varieties of plants through cross-breeding and mutagenesis techniques. In 1930 Congress enacted the Plant Patent Act (PPA) to extend patent protection to new and distinct asexually propagated (identical from generation to generation) plants. It was not until 1980 that patenting of life forms other than plants was expressly permitted.

The conditions for obtaining a plant patent are similar to those for a utility patent. An important difference, however, is that PPA provides that plant patents cannot be invalided for noncompliance with the description and enablement requirements so long as the description of the new variety is as complete as is reasonably possible. Despite the relaxed description requirements, it must still be possible to distinguish protected plants from other varieties.

The Plant Variety Protection Act (PVPA), administered by the Department of Agriculture, was enacted by Congress in 1970 to encourage the development of new sexually reproduced (combining characteristics from their parents) plants by providing an economic incentive for companies to undertake the costs and risks inherent in producing new varieties and hybrids.

Both PPA and PVPA protect only exact copies of the described plant. Although the PTO does not administer PVPA, the protection provided to breeders of new plant varieties is comparable to patent protection. PVPA includes two important exclusions to a certificate holder's protection. The research exemption precludes breeders from excluding others from using the protected variety to develop new varieties, and the farmer's exemption allows an individual whose primary occupation is growing crops for sale to save protected seed for use on their farm.

In addition to the aforementioned protections, plants may also be protected by general utility patents and specialized patents called plant patents. The ability to protect plants with utility patents was affirmed in 1985 when the PTO's Board of Appeals ruled in *ex parte Hibberd* that a corn plant containing an increased level of tryptophan was patentable subject matter. An additional benefit of utility patents in comparison to PVPA is the ability to prevent others from using a patented variety to develop new varieties. Plant patents preclude other parties from asexually reproducing, selling, or using patented plants.

A plant patent is limited to one plant or genome. A sport or mutant of a patented plant is not considered to be of the same genotype, and therefore is not be covered by the plant patent of the parent plant and requires a separate patent to protect third-party reproduction, sale, or use. An important consideration for patented plants that are grown in the wild is that if a protected plant crossbreeds with other plants or changes to produce a plant that is different from the patented type, the new plant is not protected by the original plant patent and is therefore not accorded patent protection.

As with utility patents, plant patents must be novel and nonobvious. Both the original generation and subsequent asexual reproduction of plants are considered inventive steps, so if different parties have roles in producing and propagating a novel plant, both are considered co-inventors.

PROVISIONAL PATENT APPLICATION

It is possible to file a provisional patent application up to one year prior to filing a full patent application in the United States. Provisional patent applications do not require any formal claims, oaths, declarations, or statement of novelty, but they must contain a written description of the invention, any drawings necessary to understand the invention, and the names of all the inventors. A provisional patent application must be filed within one year following the date of first sale, offer for sale, public use, or publication of an invention. The term of a patent does not include the provisional patent application period, so a patent may effectively last 21 years from the filing date of the provisional patent application.

WHEN PATENTS EXPIRE

The price of patent protection is disclosure. The best mode of practicing the invention must be disclosed in exchange for the right to exclude others from using an invention. When a patent expires, the ability to practice the invention is no longer controlled by the inventor, so disclosure can facilitate the development of competing products following patent expiration. For products like drugs, pioneers can lose a significant portion of market share overnight with the introduction of generic versions. Each day that competition can be delayed may be worth millions of dollars in revenues.

Whereas the development of new drugs requires years of research to identify and refine potential drugs and determine their safety and efficacy, generic drug development generally requires less research and development. Because generic manufacturers can avoid much of the research and development expense, they can charge significantly less than pioneers and still earn a profit. Furthermore, unlike pioneers who may only be able to make rough estimates of the potential market for a drug prior to launch, generic manufacturers can look to pioneer sales to determine market size. Generic manufacturers therefore benefit from lower development risks and relatively greater certainty of revenues (generic drugs are described in greater detail in Chapter 8).

Alternatives to using patents to exclude competitors stem from provisions of the Orphan Drug Act and the Hatch-Waxman Act. The Orphan Drug Act grants seven years of market exclusivity to pioneer developers of drugs for rare conditions, regardless of patent status. The Hatch-Waxman Act permits patent term extensions to recover time spent in clinical trials and waiting for FDA approval. These acts are described in further detail in Chapter 8.

EXTENDING PATENT PROTECTION

Because of the ability of patents to protect markets, much effort is dedicated to ensuring that patent protection is extended as long as possible. Common methods to extend patent life are to utilize market exclusivity protections granted by the FDA, and to use R&D to leverage brand strength.

FDA market exclusivity creates incentives for socially-beneficial activities which companies might otherwise not pursue. Perhaps the most popular of these incentives is the Orphan Drug Act, which provides temporary market exclusivity for drugs addressing diseases affecting small populations. The purpose of this exclusivity is to motivate the development of drugs for markets which might not otherwise provide sufficient financial merit. Additionally, companies responding to FDA requests to conduct clinical trials demonstrating that drugs are safe and effective for pediatric patients or to provide supportive evidence demonstrating that drugs can be responsibly sold and used over the counter, called an OTC-switch, are also granted temporary market exclusivity. FDA market exclusivities are described further in Chapter 8.

Another strategy to prevent generic competitors from gaining market share involves a combination of R&D and marketing. New patents can be acquired by developing new variations of drugs such as new formulations, novel combinations, new delivery methods, and gaining approval for new indications. The repeated practice of launching these line extensions is called evergreening. Because evergreening does not prohibit generic sales of drugs with expired patents, success is contingent upon consumers being willing to pay a premium for the new product rather than accepting a less expensive generic version of the original product. Some companies specialize in developing and patenting improved formulations of drugs facing expiration and selling the rights back to pioneers to give them leverage against generic competitors. R&D strategies to leverage existing products are described in Chapter 12.

In some cases, an invention may be sufficiently complicated to prevent generic production. To gain marketing approval for a generic drug, a generic drug developer must produce a drug with the same active ingredient, strength, dosage form, and route of administration as the pioneer drug, and must demonstrate bioequivalence to the pioneer product—it must have identical chemical and biological properties to the pioneer drug. Failure to produce a bioequivalent drug requires generic manufacturers to conduct costly clinical trials. An additional complication exists for protein-based drugs, where subtle differences in production conditions can yield a suboptimal product that is less effective than the pioneer version or has additional or different side effects. Whereas established guidelines exist to determine bioequivalence for small molecule drugs (traditional pharmaceutical drugs), the lack of such guidelines for protein-based drugs is a substantial hurdle for would-be generic biotechnology drug manufacturers. Generic drug approval is discussed further in Chapter 8.

The patent term adjustment provisions of the 1999 American Inventors Protection Act also provide a restoration mechanism for delays caused by the PTO. These provisions ensure that applicants still have a patent term of at least 17 years from the date of patent grant under the 20 year-from-filing patent term system. The Hatch-Waxman Act (described in Chapter 8) also allows for patent term extension for time spent waiting for FDA approval.

PATENT INFRINGEMENT, CHALLENGE, AND EXEMPTIONS

Just because a patent has been infringed does not mean that it is worth pursuing an infringement lawsuit. A common tactic of defendants in patent infringement cases is to attack the validity of the plaintiff's patent. A seven year lawsuit originating in 1992 saw Hoffmann-La Roche's rights to *Taq* polymerase—a key reagent in the Nobel Prize-winning Polymerase Chain Reaction (PCR) process that enables many important scientific procedures—revoked as the defendant, Promega Corporation, was able to convince the court of inequitable conduct in the original patent application (Hoffmann-La

Roche is the U.S. prescription drug unit of the Roche Group).

In some cases it may be better to establish a licensing or compensation agreement with an infringing party, a compromise that may be mutually preferable to a potential protracted court battle (patent licenses are discussed in the section *Patent Licensing*, below, and in Chapter 14). In the course of litigation, the plaintiff may be called upon to describe details of manufacturing processes and other privileged information in open court. Representatives of third parties may sit in on proceedings, enabling them to acquire information about the business methods of the litigating companies. Therefore, challenging potential infringers may result in the loss of both patents and trade secrets.

Out of court settlements are another approach to protect patents and prevent entry of competitors. Faced with the risks of being forced to disclose trade secrets or losing a court verdict and ceding market share to generics, out of court settlements may be a preferable method to retain intellectual property. After challenging a patent on Bayer's antibiotic Cipro, Barr Laboratories agreed to receive $25 million annually until the expiration of Cipro's patent. These settlements, known as reverse payments, are legal, although the U.S. federal trade commission and European regulators have been trying for years to have courts to rule that reverse payments violate antitrust law.

An argument in favor of reverse payments, as described in the Box *Is it Worth it for Generics to Challenge Branded Drugs?*, below, is that reverse payments are essential due to the great disparity between the relatively low cost of challenging a patent and the significant market opportunity of gaining 180 day exclusivity for a generic blockbuster. Reverse payments prevent innovators from becoming mired in endless challenges and appeals from baseless patent challenges.

CHALLENGING PATENTS

There are a variety of oversights that can lead to a patent being deemed invalid or unenforceable. The revelation of a prior publication that renders a patent obvious or not novel can lead to invalida-

Box
Is it Worth it for Generics to Challenge Branded Drugs?

The law that allows approval of generic products, the Drug Price Competition and Patent Term Restoration Act of 1984, builds in certain protections for the original drug developer (including patents and marketing exclusivities), but also allows drug sponsors of generic products to apply for FDA approval without repeating the original developer's clinical trials. The statute provides an incentive of 180 days of market exclusivity to the first generic applicant who challenges a listed patent by filing a paragraph IV certification and running the risk of having to defend a patent infringement suit.

For blockbuster drug patent, spending a few million dollars on a patent suit is not a bad investment if there is a substantial ground for challenging the patent. While it is true that basic active ingredient patents are challenged more and more frequently, no generic drug has been legally permitted to enter the US market before a branded drug's initial active ingredient patent's expiration—it is also true that patents are often challenged, found invalid, and (only) then may a generic drug enter the market.

For example, given Plavix's sales in excess of $6 billion, the promise of a 180-day exclusivity period (i.e., being the only approved generic drug) can be worth the risk of a patent challenge. In challenging the validity of a patent, all a challenger risks is the cost of the lawsuit, and they stand to gain an opportunity for hundreds of millions (perhaps billions) of dollars in revenue.

Patent invalidity suits often come down to a nuanced battle of experts trying to decide what some hypothetical person skilled in the art would or would not have found obvious a decade or two ago. The mere threat of a lawsuit claiming that the innovator's patent is invalid is similar to being held at gunpoint, since a jury's ruling decision that the patent invalid would leave them with no way to protect a drug worth billions of dollars in annual revenues.

Therefore, the drug company will often decide to resolve the dispute out of court, with the generic companies agreeing to give up their claims in exchange for cash settlements. The generic versions of the drugs then enter the market when the patents expire. But, the alternative would be for the companies to continue legal battles through endless appeals.

Contributed by Stephen Albainy-Jenei of Frost Brown Todd LLC: www.patentbaristas.com

tion. Falsification of data or failure to disclose the best method to practice a patent can also invalidate a patent. An inventor who has not assigned his/her rights to an invention may practice the patented invention and may grant licenses to third parties without the permission of the co-inventors. The sponsoring institution that funded research leading to a patent may also have claim to partial ownership of an invention. In addition, all inventors must usually join as plaintiffs in an infringement suit. By being excluded from an infringement suit, a co-inventor can prevent the enforcement of a patent. Therefore, failure to list all the inventors of a patent can frustrate licensing agreements and potentially render a patent unenforceable.

While the U.S. Patent and Trademark Office is responsible for granting patents, the courts have the final determination of patent validity. Patents can be invalidated based on technical or legal grounds. Proof of failure to satisfy the requirements of novelty, non-obviousness, or utility can render a patent invalid.

An out-of-court mechanism to challenge the validity of a patent is to request a reexamination by the PTO. Reexamination can only be initiated on the basis of invalidity due to prior art; not falsification of data, lack of enablement, or other challenges. Prior art revelations sufficient to question the validity of the patent were traditionally followed by an *ex parte* procedure involving only the patent holder and PTO. In these cases, the inability of the challenger to respond to the patent holder's defenses often resulted in the patent holder redrafting their patent with stronger claims, an outcome at odds with the purpose of reexamination. In response to criticism of *ex parte* reexamination, the PTO introduced an optional *inter partes* reexamination procedure on November 29th 2000, applicable to patents filed on or after November 29th 1999. A limitation of *inter partes* reexamination is that the requester is prevented from using the same prior art arguments in subsequent infringement litigation.[2]

The only other means to challenge a patent is through litigation. The Supreme Court's 2007 decision in *MedImmune v. Genentech* (see Box *MedImmune v. Genentech: Licensees gain power* in Chapter 14)

---

2 Derzko, N.M., Behringer, J.W. Inter partes reexamination: a potentially useful approach to challenging invalid biotechnology patents. *Nature Biotechnology*, 2003. 21(7):823-825.

Box

Amgen v. Transkaryotic Therapies: Strategic patenting

Amgen's Epogen, the first billion dollar biotechnology drug, is produced by introducing human genes into hamster cells. Transkaryotic Therapies' (since acquired by Shire) Dynepo promotes production of erythropoietin from cultured human cells by stimulating expression of the existing erythropoietin gene rather than introducing the gene. Production differences aside, both drugs stimulate proliferation of red blood cells in patients suffering from kidney failure.

Amgen filed its first patent application on the recombinant erythropoietin gene in 1983, obtained its first patent on EPO in 1986, and then continued to develop more extensive patent protection. To protect Epogen, Amgen filed a series of divisional patent applications stemming from their original 1983 patent. Divisional patent applications are employed when an application makes too many claims to be considered a single patent, covering the same specification as the parent patent but claiming different inventions.

In 1997 Amgen filed suit alleging that Transkaryotic Therapies and partner Aventis SA were infringing on five of Amgen's EPO patents, all based upon the initial 1983 patent application filing. Transkaryotic Therapies' countered, arguing that their EPO production strategy was different from Amgen's patented method and that Amgen's patent claims were legally insufficient and therefore invalid.

Amgen ultimately succeeded because of their series of patents protecting different elements of Epogen production. These patents were targeted specifically at Transkaryotic Therapies' defense case. The key to this strategy was demonstrating that new claims were clearly contemplated and described in the initial application.[1]

---

1 Williams, K.M., What lessons should a biotechnology company take from the Amgen v. TKT patent infringement suit? Palmer & Dodge LLP, 2000.

---

altered the requirement for parties wishing to directly challenge a patent. Previously patents could only be challenged by parties accused of infringement or in fear of being accused of infringement. A company wishing to challenge a patent needed to knowingly infringe the patent and await infringement proceedings, risking trebled damages for willful infringement if they lost the challenge. In the wake of *MedImmune v. Genentech,* challengers may now pay royalties in escrow, avoiding trebled damages, while challenging a patent.

A downside of resorting to patent challenge is that biotechnology-related patent litigation is very expensive, typically costing $3-10 million for each party.[3] To protect individual patents and discourage challenge, drug developers often file multiple patents on various aspects of individual products, avoiding complete loss of monopoly in the event that a single patent is invalidated. This can also be a good strategy to pursue infringers, as demonstrated by the case of *Amgen Inc. v. Transkaryotic Therapies Inc.*

PATENT EXEMPTIONS

A motivation for the government to provide exclusive rights to practice an invention through patent grant is to provide the public with new and useful inventions by providing the inventor an incentive to publish the best mode for practicing an invention. The inventor's rights must therefore be limited in cases where the benefit to the public outweighs the harm to the inventor. Two such inventor limitations are covered by the experimental use and government infringement exceptions.

EXPERIMENTAL USE

The experimental use exception holds that a patent is not infringed if the use is limited to research or experimentation and the user does not profit from the experimental use. According to the PTO,

> A use or sale is experimental ... if it represents a *bona fide* effort to perfect the invention or to ascertain whether it will answer its intended purpose. If any commercial exploitation does occur, it must be

3 Apple, T. The coming US patent opposition. *Nature Biotechnology*, 2005. 23(2):245-247.

merely incidental to the primary purpose of the experimentation to perfect the invention.

A limitation of the experimental use exception is that the courts have interpreted it rather narrowly. In the 1984 case of *Roche Products, Inc. v. Bolar Pharmaceutical Company*, Roche was seeking damages for Bolar's tests seeking FDA approval of a product for which the patent would soon expire. Although Bolar stated no intention to produce the drug for sale until after the patent expired, the court held that the use of the patented product was not within the experimental use exception.

The Hatch-Waxman Act (described in Chapter 8) was enacted to accommodate generic drug testing prior to patent expiration, providing a limited research use exemption: "It shall not be an act of infringement to make, use, offer to sell, or sell [a patented invention] ... solely for uses reasonably related to the development and submission of information under a federal law which regulates the manufacture, use, or sale of drugs or veterinary biological products."

A notable omission in the language of this exemption, known as the Hatch-Waxman safe harbor, is that it does not explicitly state that the exemption only applies to *generic* drug development, establishing the possibility that the exemption might enable all drug developers to use patented tools without license for new drug development as well. A series of patent cases examined the limitations of the safe harbor and saw the courts generously extending the research exemption to any activity related to drug approval, for generic and new drugs alike.[4]

A key ruling on the scope of safe harbor protection emerged from the case of *Integra LifeSciences Holdings Corp v. Merck KGaA*. In this case, Integra sued Merck and other parties for infringing a group of patents describing protein sequences that promote or inhibit cell adhesion. Merck had been using the patented protein sequences to develop new drug candidates for a variety of diseases. Upon learning of Merck's activities, and failing to convince Merck to license the technology, Integra sued Merck for patent infringe-

---

4 Raubicheck, C. *et al.* Integra v. Merck: a mixed bag for research tools. *Nature Biotechnology*, 2003. 9(21):1099-1101.

ment. Although the U.S. Court of Appeals for the Federal Circuit initially ruled in Integra's favor, the Supreme Court unanimously reversed the ruling in 2005, clearly indicating that broad immunity from patent infringement exists for pre-clinical research and experimentation "reasonably related" to the process of developing information required for submission to the FDA.

The Supreme Court did not specifically state that Merck's activities qualified for safe harbor protection, but argued that the Federal Circuit's interpretation of the exemption to not include research which might not ultimately be submitted to the FDA was inconsistent with the language of the exemption. The implication of the Supreme Court ruling is that pharmacogenomics tools such as diagnostic tests and single nucleotide polymorphisms (see Chapter 5) can potentially be used for drug development without infringement. It is important to note that while the current interpretation permits research on patented compounds, commercialization of drugs containing patented elements is not possible without license.

Government Use

The U.S. government is exempt from patent infringement. The rationale for this position is that the government acts for the public good and therefore its interests, being those of the public, take precedence over private rights such as patent rights. A 1910 statute provides patent owners the right to reasonable compensation for the use of a patented invention by the government without license. This exemption was widened in 1918 in response to concerns of the government's ability to obtain and produce sufficient quantities of war-related supplies. Congress amended the law to cover not only items manufactured by the United States, but also those used or manufactured by, or for, the United States. This amendment renders contractors free from liability in infringement of patents to manufacture anything for the government and dictates that the inventor should seek reasonable compensation from the government, not the contractor.

Recipients of grant funding from U.S. government agencies may satisfy the definition of government contractors. Following enactment of the Bayh-Dole Act in 1980, universities were granted the right to license patents emerging from federally funded research (see *Bayh-Dole Act* in Chapter 9). The establishment of means for academic researchers to commercialize their discoveries blurred the distinction between academic and commercial research, potentially enabling academic researchers to infringe patents for profit without penalty.

A 1992 court battle between Roche and Promega over the unlicensed sale of a patented enzyme challenged application of the experimental use exemption by academic researchers. Roche, the plaintiff, claimed that scientists are "in the business of doing research in order to ... attract private and government funding through the publication of their experiments in the scientific literature, create patentable inventions, and generate royalty income for themselves." A separate trial invalidated Roche's patent, preempting the question of experimental use and leaving Roche's challenge unanswered.

A definitive statement on the application of the experimental use exemption within academic settings was made in 2002 when the U.S. Court of Appeals for the Federal Circuit ruled in *Madey v. Duke* that university scientists can be sued if they infringe a patent while conducting research. The case stems from a laser invented and patented by John Madey while he ran a laboratory at Duke University. After he had left for the University of Hawaii, Duke researchers continued to use his laser, so Madey sued for infringement. The judges affirmed that activities that are "for amusement, to satisfy idle curiosity, or for strictly philosophical inquiry" are exempt from patent infringement. While no direct profit was derived from the use of Madey's patented laser, Madey's attorneys successfully argued that Duke researchers advanced their "legitimate business objectives" to obtain grants, attract faculty, and such, and were therefore not exempt. Accordingly, the use of research tools to advance business objectives, regardless of profitability, is subject to liability for patent infringement.

PATENT LICENSING

Licenses enable parties to obtain the right to use patented technologies. Licensees often pay up-front payments on signing the agreement and maintenance or royalty payments over the life of the license. Licensees may also be tasked with policing infringers. This section focuses on legal considerations on licensing patents. License types, licensee and licensor responsibilities, and commercial considerations of patent licensing are described in greater detail in Chapter 14.

STACKING ROYALTIES AND SUBMARINE PATENTS

The number of patents on applications of genes and biotechnology processes represents a significant burden for some companies. Submarine patents exacerbate this concern. Patents that emerge after other companies have unknowingly infringed them are termed submarine patents. Submarine patents can emerge from new applications or from divisional applications that expand on an existing patent. Some countries, such as those in Europe, make the content of patent applications public before they are granted, effectively blocking one source of submarine patents. In the United States patent applications traditionally remained secret until a patent was issued. As of November 2000 all U.S. patents are published 18 months after being filed, with the exception of provisional applications, abandoned patents, and applications subject to secrecy orders. Inventors may also prevent pre-approval publication by certifying that they are not seeking foreign patents on the same invention.

One proposed solution to ease the administrative and financial burden of managing multiple licenses is to form patent pools. A patent pool is an agreement between one or more patent owners to license their patents to one another or third parties. In 1917, an aircraft patent pool included virtually all the aircraft manufacturers in the United States. Patents essential to the MPEG-2 multimedia compression technology standard are a more recent example of a patent pool. A key element in a successful patent pool is the adoption of standards among the pool participants, facilitating adoption by licensees and creating a competitive advantage for licensees to utilize the pool instead of independently aggregating licenses.

Another way to manage licensing issues is for a group of companies to collaborate and develop public domain projects. The prospect of patents limiting commercial application of single nucleotide poly-

---

Box

Monsanto hit by $185 million torpedo

On February 27th 2006, the very day a patent infringement lawsuit was scheduled to start, Monsanto announced an agreement to pay the University of California a $100 million up-front royalty to resolve issues regarding possible infringement and validity of patent rights relating to the production of bovine somatotropin (BST). In addition to the up-front royalty payment, Monsanto agreed to pay the university 15 cents per dose of BST sold to dairy producers, with a minimum annual royalty of $5 million.

BST, introduced to the market by Monsanto in 1994, increases milk production in cows by 8 to 12 pounds a day. The central patent in this case was filed in 1990 as a continuation of applications dating back to 1980 and describes a DNA sequence coding for bovine growth hormone and an expression system to produce the hormone. This patent did not issue until February 17th 2004, a decade after Monsanto first marketed BST.

While the central patent in this case expires in 2021, a second patent claiming BST as a product of the recombinant DNA sequence expires in 2023. Adding Monsanto's minimum $5 million annual royalty to the $100 million up-front payment brings Monsanto's liability to at least $185 million.

---

morphisms (SNPs), which can be used as disease markers, prompted several major pharmaceutical companies to create the SNP Consortium. The SNP consortium identified SNPs and placed them in the public domain so that no single party could inhibit their use.

REACH-THROUGH CLAIMS

The objective of patenting an enabling technology is to control its use by others. While some patents are primarily directed at ensuring freedom to operate or preventing other companies from practicing an invention, others describe processes that can be sold or licensed. Licensors have traditionally employed "reach-through" claims to preemptively obtain a preferential starting position in negotiations over the scope of licensing rights, claiming rights to royalties from, or rights to use, drugs or other physical or intellectual property produced using patented processes within a patent itself.

A case example of reach-through claims delivering significant royalties is in the patents on the Cohen-Boyer gene splicing techniques that arguably spawned the biotechnology industry (see *Start-up* in Chapter 10 and Box *Cohen-Boyer: Broad licensing to maximize revenues* in Chapter 14). These patents covered the fundamental gene splicing technology and reached-through to claim recombinant organisms created using the technology. Licenses for these patents earned nearly $300 million in royalties.

Reach-through claims often run afoul of enablement requirements. To patent an invention, the inventor must provide a written description that describes the invention in sufficient detail to enable a person with expertise in the relevant field to practice the invention without "undue experimentation." In meeting this requirement, it is not sufficient for reach-through patent claims to describe the features of inventions stemming from a patent; the claims must describe the very structure of these inventions.

It is not uncommon for early patents in emergent fields, like the Cohen-Boyer gene splicing patents, to be granted reach-through claims, but as fields mature, reach-through claims are less common. Early biotechnology patents requested protection for "prophetic" claims; untested inventions that are described in sufficient detail to enable a skilled worker to practice the invention. The 1991 case of *Amgen Inc. v. Chugai Pharmaceuticals and Genetics Institute* led the U.S. court of appeals to clarify the requirements for a written description in patent applications. Chugai and Genetics Institute were sued for infringing Amgen's claim to a DNA sequence encoding human erythropoietin. The court decided that it was not sufficient to know how a compound of an unknown structure *might* be isolated; the inventor had to actually practice the invention. Subsequent lawsuits further defined the requirements for invention, specifying that patents must describe an invention and not merely define its characteristics.

In a similar case, Chiron sued Genentech over sales of Herceptin, a drug with annual sales exceeding $1 billion. Herceptin is a humanized monoclonal antibody which binds the HER2 protein found on certain cancerous cells. Chiron asserted that Herceptin infringed their 1984 patent, which claimed a mouse monoclonal antibody binding to HER2. Chiron's claims to Herceptin were rejected because their patent was insufficient to enable discovery of the drug. The courts found in Genentech's favor because Herceptin was a humanized antibody, and Chiron's patents made no claims of humanized antibodies. Humanized antibodies did not emerge until several years after the patent application, and required significant experimentation to produce.

Another high profile case involved the University of Rochester's claim to an emergent class of painkillers. In the 1990s researchers at the University of Rochester found that inhibiting the Cox-2 enzyme, while not affecting the closely related Cox-1 enzyme, reduced inflammatory pain without the side effects associated with drugs inhibiting both enzymes. They filed a patent covering selective inhibition of the Cox-2 enzyme, but did not include any claims specifically describing drugs exhibiting this activity. When G.D. Searle (later acquired by Pfizer) developed Celebrex, a Cox-2 inhibitor, the University of Rochester sued them for patent infringement. The patent appeals court affirmed the invalidity of reach-through claims covering drugs specifically inhibiting Cox-2 in 2004, arguing that the University of Rochester's patent did not satisfy the requirements of patentability by failing to describe any such drug or how to make such a drug.

TRADE SECRETS

An alternative to patenting inventions is to retain them as trade secrets. A trade secret is an item of information—a customer list, business plan, or manufacturing process—that has commercial value and is not exposed to the public by the holder. There are two basic types of trade secrets. Technical secrets refer to research and development methods and tools; business secrets refer to items such as marketing, sales, financial, and administrative data. A distinction is also made between an employee's abilities and a company's proprietary knowledge. A scientist's research expertise does not qualify as a trade secret, but a company's unique research techniques and methods may qualify.

Unlike the limited term of patent protection, trade secrets can potentially last indefinitely. A limitation of this indefinite term is that the holder of a trade secret has a responsibility to make efforts to protect it. Courts may refuse to recognize a trade secret if reasonable efforts have not been made to keep the information from being disclosed.

A disadvantage of trade secrets, when compared to patents, is that they do not grant the right to exclude others from practicing an invention. If a competitor is able to independently develop or reverse-engineer an invention, or if the owner of the secret accidentally makes it public, then the owner has little recourse for preventing use.

Trade secrets are best used for processes which are not subject to reverse engineering, such as in-house manufacturing, drug-screening and related techniques, and compiled computer programs, where control of the protected invention is maintained by the owner. Processes not subject to patent protection, perhaps due to little-known prior art, are also good candidates for trade secret protection. Processes for which patents would be difficult to enforce, such as drug screening on molecular evolution (see Box *Licensing technology or keeping it in-house* in Chapter 14) may also be best protected as trade secrets.

A significant disadvantage of patents, relative to trade secrets, is that patents eventually expire and require disclosure of the best mode to practice an invention, which may necessitate the publication of trade secrets involved in production processes. Furthermore, because it is not possible to anticipate all the possible applications of an invention, patent publication may enable the development of non-infringing competing products. In deciding how best to protect an invention these significant differences necessitate weighing the near-term protective effects of patents against the long-term benefits of trade secrets.

TRADEMARK

A trademark is a word, symbol, or phrase used to identify a particular manufacturer or seller and their products, and to distinguish them from others. Unlike copyrights or patents, trademark rights can last indefinitely if the owner continues to use the mark to identify its goods or services. Trademarks can also extend to the look and feel of a product or logo, provided it does not confer any sort of functional or competitive advantage. Drug companies trademark the names and physical characteristics of their drugs, which is why generic drugs have different names and look different from pioneer products.

Trademark protection is granted to the first company to use a mark. First use can be proven by either providing evidence of use or by registering the mark with the PTO. Without registration, the scope of trademark protection is limited to the geographic area in which the marked product is sold. A mark that has been registered with the PTO is granted for use nationwide, limited to the extent that others are already using the mark within a specific area.

The rights to a trademark can be lost through abandonment, improper licensing or assignment, or genericity. Non-use of a trademark for three consecutive years is considered evidence of abandonment. One reason why companies are very particular about the representation of their trademarks is that a trademark licensed without adequate quality control or supervision by the owner may be canceled. The rationalization in such situations is that the trademark no longer identifies the goods of a particular provider. Trademarks that are originally unique can also become generic over time. A trademarked word is considered generic when, in the minds of a substantial majority of the public, the word denotes a type of product and not a specific source or manufacturer. In trademark cases, courts look to dictionary definitions, use of the term in newspapers, and evidence of attempts by the trademark owner to police the mark.

A relevant example of the loss of a trademark to genericity is Bayer's loss of its Aspirin trademark. Because the word *aspirin* was consistently used without the product descriptor *pain reliever* following it, aspirin became a generic term for pain reliever and was

---

Box
Johnson & Johnson sues Red Cross over trademark

Unlike patents or copyrights, the term for trademarks is unrestricted. In order to maintain a trademark over its unlimited lifespan, trademark holders must actively enforce their marks. Failure to do so can be construed as abandonment, which can lead to loss of protection.

In 2007 Johnson & Johnson found themselves in the uncomfortable position of suing the Red Cross for trademark violation. Johnson & Johnson and the Red Cross had shared use of the red cross symbol for over a century, with Johnson & Johnson using it on commercial products and the Red Cross using it in relation to its humanitarian relief efforts. According to the lawsuit, the Red Cross started licensing the red cross symbol for use in commercial products such as humidifiers, medical examination gloves, bandages, nail clippers, combs, and toothbrushes—many of the same products for which Johnson & Johnson used the same symbol.

The Red Cross claimed that this use was consistent with its mission, as the symbol was being used on first-aid and "disaster" kits. Johnson & Johnson argued that this commercial use was in violation of the terms of their agreement and, while not wishing to refuse the Red Cross' use of the symbol for humanitarian objectives, felt compelled to restrict the Red Cross from licensing the red cross symbol in order to protect their own use of the trademark.

Following an initial ruling which sided with the Red Cross, deciding that their Congressional charter gave them the right to use the symbol even for business purposes, in June 2008 both parties agreed to dismiss their remaining claims and counterclaims.

---

therefore no longer protected under trademark law.

COPYRIGHT

Copyright protection exists for "original works of authorship fixed in a tangible medium of expression, now known or later developed, from which they can be perceived, reproduced, or otherwise communicated"—the material must be original and subject to communication.

Naturally occurring DNA sequences are ineligible for copyright protection because they lack originality. The general argument for copyrighting a novel DNA sequence is that the act of creating a new sequence of DNA constitutes an original expression. As an analogy, computer programs exist as magnetic patterns on computer disks, yet they can also be expressed on paper, and therefore may be protected by copyright. DNA sequences, like other forms of information, can also be expressed on paper despite their existence in more complex forms, and may therefore be eligible for copyright protection. Unfortunately, this analogy does not extend copyright protection to DNA sequences because of the utility of DNA sequences, which precludes copyright eligibility. Copyright law protects the expression of ideas, but cannot be used to prevent practice of ideas.

Although one might hypothetically avoid the utility restriction by copyrighting only some of the numerous DNA sequences that can be used for a given application, this would be of little practical use. While there is some variety in the DNA sequences that can be used in any application, it would be necessary to copyright them all in order to effectively control usage. There is no practical value in copyrighting only a specific DNA sequence if others can be used in its place.

Therefore, the only copyright-protectable DNA sequences are those that the copyrighter has no foreknowledge of and are not useful, satisfying the originality and lack of utility requirements. Such DNA sequences have little utility in biotechnology.[5]

THE ROLE OF INTELLECTUAL PROPERTY IN BIOTECHNOLOGY

This chapter opened with a quotation that addresses a common misconception in patenting biotechnology inventions. It is not possible to patent genes; only applications of genes may be patented. Biotechnology research is focused on discovering and enabling applications of molecular biology. The emphasis on developing and commercializing novel products and services means that intellectual

5 Silva, J.G., Copyright protection of biotechnology works: Into the dustbin of history. *Boston College Intellectual Property and Technology Forum*, 2000.

Box

The SNP Consortium

Recognizing the value of single nucleotide polymorphisms (SNPs) in screening for disease predisposition, predicting drug response, and assisting research, 11 major pharmaceutical companies and the Wellcome Trust formed the SNP Consortium in 1998 with the goal of mapping SNPs and putting them into a freely-available database. SNPs represent discrete genetic differences that can enable rapid identification of an individual's susceptibility to specific diseases or likelihood to benefit or suffer side effects from certain drugs. SNPs are discussed in further detail in chapter 5.

The motivations to form the SNP Consortium were to share the cost of identifying SNPs through collaboration rather than having interested parties duplicate each other's efforts, and to make identified SNPs publicly available to prevent individual parties from securing exclusive rights. In order to establish a date of discovery and prevent subsequent patenting, consortium members filed patent applications on identified SNPs, but agreed not to allow any patents to issue.

The SNP mapping project, expected to develop up to 300,000 SNPs, started in April 1999 and was anticipated to continue until the end of 2001. In the end many more SNPs were discovered—1.5 million in total—than were originally envisioned.

---

property protection is essential to foster innovation in the biotechnology industry. Patents and other forms of intellectual property enable companies to attract financing and endure lengthy development efforts by providing market exclusivity for commercial applications.

The benefits of market exclusivity are countered by concerns that patents impede scientific progress and limit access to useful tools and products. While intellectual property protection does enable companies to ask relatively higher prices for protected products, it is important to consider the basis of intellectual property protection.

Ideally, the purpose of intellectual property protection is to enable and accelerate development of inventions that would not occur without protection. Substantial investments are required to produce innovative products and services. In drug development, for example, the bulk of effort is spent identifying potential drugs and proving them to be safe and effective. Many years and hundreds of millions of dollars are required to develop and demonstrate the safety and efficacy of even a single drug.

The sophistication of research tools means that it is often possible to reverse-engineer a proven drug or other biotechnology-derived product. Without patent protection, competitors would be free to copy successful products at a much lower development cost than pioneers, and could therefore profitably sell them at lower prices. This premature price competition would prevent pioneers from recouping their research and development investments, favoring a commodity-based rather than an innovation-based industry model. This is exactly what was witnessed in China and India before their patent laws were strengthened to protect drugs; these countries had thriving generic drug industries with strong exports, but little inward investment or research and development activity.

To avoid the unwelcome situation in which competitors prevent pioneers from securing compensation for their efforts, research is focused on applications that can be protected by patents or other means. The scope of what can be protected thereby indirectly defines which applications biotechnology companies will develop.

In addition to protecting intellectual property and establishing a virtual monopoly, patents also have some indirect benefits. During development, in the absence of profits, patents can imply the ability to generate profits. The demonstrated ability to turn ideas into inventions can attract investors and development partners. Revenues can also be derived from the sale or license of patents.

Concerns over restriction of research due to an overwhelming number of patents are somewhat resolved by the decrease in patent value as the number of patents increases; the price of patents must reflect their value. A single high priced patent that controls an area of research may also encourage competitors to develop alternative methods and license them at lower prices. Patents, like physical property, are subject to the effects of supply, demand, and competition.

The requirements for novelty and non-obviousness dynamically reflect the current state of scientific knowledge. Furthermore, in order to be granted a patent, an inventor must disclose publicly how to best practice the invention. Therefore, patents provide a means by which the public can gain valuable cutting-edge scientific knowledge and abilities in exchange for a temporary grant of monopoly, which allows innovators to recoup their investments in research and development.

Chapter 8

Regulation

> Perhaps the most important discovery of the twentieth century
> was to learn to identify and read the code of life. And perhaps
> the greatest challenge we will face in the twenty-first century
> … is how … and when … to apply this knowledge.
> *Juan Enriquez, Harvard Business School*

Regulatory approval is the second most important measure of the quality of a product from an investment perspective, after patent strength. Regulations ensure that drugs are safe, effective, and appropriately labeled, and that biotechnology products such as plants and bioremediating bacteria will not harm the environment. In addition to protecting public health, regulations also enable objective measures of product efficacy and safety, facilitating their comparison.

Regulations serve the dual functions of setting limits on the applications of biotechnology and providing incentives for innovation. An example of an regulation which drives innovation, the Orphan Drug Act, provides incentives for companies to produce treatments

Table 8.1 *Biotechnology regulating bodies*

Food and Drug Administration	Food, feed additives, veterinary drugs, human drugs, and medical devices
Department of Agriculture	Plant pests, plants, and veterinary biologics
Environmental Protection Agency	Microbial and plant pesticides of chemical and biological origin, new uses of existing pesticides, and novel organisms that may have industrial uses

Table 8.2 *Regulation of biotechnology products*

Food	
Plant products	FDA
Animal products	FDA, USDA
Food additives	FDA
Dietary supplements	FDA
Animal feed	FDA
Drugs	
Human synthetics	FDA-CDER
Human biologics	FDA-CBER
Animal synthetics	FDA
Animal biologics	USDA
Industrial Products	
Cosmetics	FDA
Pesticides	EPA
Other substances, if toxic	EPA

for rare disorders. These incentives aim to encourage research into areas that might otherwise not demonstrate sufficient profit potential to merit investigation.

The agencies primarily responsible for regulating biotechnology in the United States are the Food and Drug Administration, Department of Agriculture, and Environmental Protection Agency. Products are regulated according to their intended use. Some products are regulated by more than one agency.

FOOD AND DRUG ADMINISTRATION

The FDA is responsible for regulating food, feed additives, veterinary drugs, human drugs and medical devices. Drug makers must test and gather data showing whether a drug is stable at certain temperatures and in powder, injectable, pill, or tablet form, and whether it can be manufactured repeatedly with consistent quality. In addition to defining a development path by which to demonstrate the safety and efficacy of drugs, clinical trials also play a vital role as valuation milestones (see Box *Clinical trials provide valuation milestones* in Chapter 12).

The FDA also regulates manufacturing and delivery processes.

Good manufacturing practice guidelines ensure the quality and purity of chemical products that are intended for use in pharmaceutical applications and describe the controls to ensure that the methods and facilities used for production, processing, packing, and storage result in drugs with consistent and sufficient quality, purity, and activity.

The Prescription Drug User Fee Act (PDUFA) establishes three types of user fees: application fees, establishment fees, and product fees. User fee reductions and waivers are available for eligible small companies, enabling the FDA to benefit from user fees while avoiding an unnecessary burden on companies with limited economic resources.

Small drug development firms, defined by the FDA as those with less than 500 employees and are not a subsidiary of a larger business, are granted a reduction and deferral of application fees, provided they do not currently market an approved product. Establishment and product fees may also be waived for any business that meets specific requirements. PDUFA also permits waivers or reductions of user fees if it is necessary to protect public health, if the fees present a significant barrier to innovation, if the fees exceed the projected FDA cost of reviewing an application, or if a product contains the same active ingredient filed for by another applicant for which user fees were not levied.

The FDA's role as arbiter of which drugs may be sold, and for which indications they can be marketed, places them at the center of controversy between demand for prompt access to vital medicines and demand for medicines with rigorous evidence of safety. The agency is often simultaneously blamed for impeding translation of drugs from labs to the marketplace, and for inadequately regulating the safety of drugs. The two objectives are often at odds with each other. Ensuring safety requires extensive testing, and extensive testing means longer periods spent seeking approval.

CLINICAL TRIALS

Biotechnology drug development is covered in detail in Chapter 4. The clinical trial process is described here. The basic process for drug development is as follows:

1. Potential drug compounds are identified, using laboratory and animal models.
2. Potential drugs are tested in animals and cell cultures to determine if they are safe enough for clinical trials in humans. These pre-clinical tests attempt to predict the ways in which drugs may interact with the human body.
3. If a drug passes pre-clinical development, human clinical trials follow to ultimately provide information, necessary for FDA approval, on its safety and efficacy.

On average, one compound in a thousand will make it to clinical trials. Roughly 70 percent of drugs that complete clinical trials receive FDA approval.

To pursue human studies, a sponsor must first submit an Investigational New Drug (IND) application to the FDA for approval. The IND must contain sufficient information for the FDA to justify testing a drug in humans. Approximately 85 percent of all IND applications move on to clinical trials. Because larger drug companies can more easily afford expensive human trials, smaller companies often form alliances with established partners to fund the process.

Clinical trials are conducted to determine the safety and efficacy of drugs. Drugs must be proven safe and effective before FDA approval for marketing can be granted. While safety is the primary concern, a drug with detrimental side effects may be acceptable if there are no better treatments and the severity of disease warrants it. Most companies file for and receive patents for the commercial use of compounds during pre-clinical development. Much of a drug patent's life can therefore lapse during clinical trials and while waiting for regulatory review. The Hatch-Waxman Act, described later in this chapter, contains provisions to partially recover time spent in clinical trials and waiting for FDA approval.

Every clinical trial in the United States must be monitored by an Institutional Review Board (IRB). An IRB is an independent committee of physicians, statisticians, community advocates, and others that ensures that the risks are as low as possible and are worth any potential benefits, and that the rights of study subjects are protected.

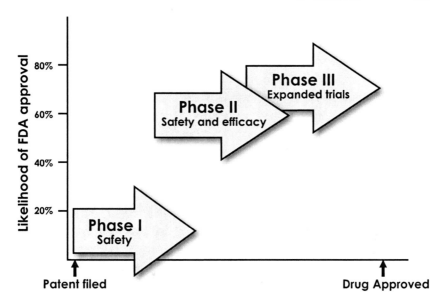

Figure 8.1 *Clinical trial phases*

IRBs must contain at least five experts and lay people of both sexes with varying backgrounds. Consultants may be retained to assist in the review of subjects requiring specialized knowledge not held by IRB members. In addition to reviewing scientific activities, an IRB must also be able to ascertain the acceptability of applications and proposals in terms of institutional commitments and regulations, applicable laws, standards of professional conduct and practice, and community attitudes.

PHASE I

While the purpose of clinical trials is to determine the safety and efficacy of a drug, the primary consideration is the safety of the participants. Phase I trials are designed to determine the safety of drugs. These trials involve a small number of healthy volunteers or affected patients who are given doses ranging from sub-clinical to potentially toxic.

To minimize risk to human subjects, all drugs must undergo extensive pre-clinical development to determine their effects on animals, and predicted effects in humans, prior to Phase I trials.

Beginning with human trials in Phase I, drugs must be produced under current good manufacturing practices (cGMP). To satisfy FDA cGMP guidelines, manufacturers must be able to demonstrate compliance with regard to facilities, raw materials handling, and manufacturing control and associated documentation (see *Manufacturing* in Chapter 5).

There are two basic types of Phase I trials. *First-in-man* studies are primarily concerned with establishing the safety of a compound. These studies start with an initial small dose that is given to a small group of participants. If no adverse effects are seen, escalating doses are given to new groups of participants. Dose limiting toxicity is observed when the dose is escalated to the point that dangerous side effects are seen. The other type of Phase I study is the *clinical pharmacology* study. The goal of this study is to determine the pharmacokinetics of a compound: how a drug is absorbed, distributed in the body, metabolized, and excreted.

Data from Phase I trials are essential for the design of appropriate Phase II trials. Taking shortcuts in Phase I trials may only serve to see a compound fail in more expensive Phase II or Phase III trials.

Phase I testing ranges from one to three years on average. Historically, drugs in Phase I have a 10 percent chance of making it to market. If Phase I trials do not reveal unacceptable toxicity, a drug can proceed to Phase II testing. While failure in a phase I trial indicates that the tested form of a drug is unacceptable, success may still be possible by modifying a compound based on observed data.

PHASE II

The emphasis in Phase I trials is on safety; Phase II trials introduce effectiveness. Phase II trials consist of small, well-controlled experiments to further evaluate a drug's safety, assess side effects, and establish dosage guidelines. Drugs are given to volunteers (usually between 100 and 300 patients) who actually suffer from the disease or condition being targeted by a drug.

This phase is where the minimum effective dose, maximum tolerable dose, and optimum dosage of drugs are established. Drug regimens are tested to see how often a drug must be administered; a drug may be effective if taken once a month, or may require administration several times a day. Statistical end points are established for drugs, representing the targeted favorable outcome of the study. The current standard of care for a medical condition can be used as a benchmark in setting the end point.

Phase II trials last an average of two years. If Phase II trials indicate effectiveness, a drug can proceed to Phase III trials. A drug that moves on to Phase III testing has an approximately 60 percent chance of being approved by the FDA. A properly designed and administered Phase II trial can help select dosage regimens and treatment indications that make Phase III trials faster and easier. Rushing this process may require repetition of Phase III trials or lead to outright failure.

It is important to note that successful Phase II trials results may not necessarily be followed with Phase III trials. A 2007 study of 200 phase II trials with "encouraging results" for various cancers found that only 13 percent progressed to Phase III trials. Some of the impediments blamed for lack of progression were insufficient finances and an inability to recruit sufficient patients for Phase II trials. It was also suggested that companies may use Phase II trials to promote off-label use—prescription for an unapproved indication—instead of using larger trials and formally seeking FDA approval for additional indications.[1]

Phase IIA / IIB

Because clinical trial phases can take years to complete and often have multiple objectives, drug developers have taken to sub-dividing the phases in order to express a sense of progression. This division is most prevalent in Phase II trials, where safety data from Phase I trials are confirmed and expanded, and dosage and administration profiles are established in preparation for Phase III trials.

While the terms Phase IIa and Phase IIb are not recognized by the FDA, they are used useful devices to convey a drug's position in the approval process to investors, analysts, and partners. As

---

1 Mundell, E.J. Study questions dead-end cancer clinical trials. *HealthDay.* September 10, 2007.

Box
Erbitux: Poor study design

ImClone's Erbitux originally failed to achieve a biologics license approval due to poor study design in a high-profile spectacle that saw ImClone's CEO Sam Waksal and media magnate Martha Stewart imprisoned amid allegations of insider trading. Erbitux was eventually approved, but had it failed outright it would have cost senior partner Bristol-Myers Squibb nearly $2 billion, making it an excellent case study of drug development gone wrong.

Erbitux is a monoclonal antibody that works by binding and inhibiting receptors that ordinarily signal cells to divide, preventing the proliferation of cancer cells. ImClone filed for an IND for Erbitux in 1994. ImClone researchers examined the safety and efficacy of the drug for a variety of cancers, including as a treatment in combination with a standard chemotherapy drug, irinotecan, in refractory colorectal cancer.

A key Phase II trial enlisted 139 patients and measured the effectiveness of Erbitux in combination with irinotecan in reducing tumor size—a surrogate endpoint for patient survival—and was intended to examine the effectiveness of the combination for third-line therapy for patients with metastatic cancer who had failed to respond to irinotecan alone. While this trial was not intended to be used as a pivotal trial for approval, encouraging results led ImClone's management to file a biologic license application (BLA) on Erbitux earlier than planned. In a meeting with the FDA, ImClone and the FDA agreed that in order for a BLA to be granted, (1) patients in the study must have tumors that progressed despite prior treatment with irinotecan; (2) at least 15 percent of the patients must respond to the combined regimen of Erbitux and irinotecan, with at least a 50 percent reduction in tumor size; and (3) the findings must meet statistical requirements.

In granting fast track status to Erbitux, the FDA required ImClone to conduct a small-scale study of Erbitux as a single agent (i.e., not in combination with irinotecan) for patients with colorectal cancer refractory to irinotecan, to set a baseline for the combination therapy and to demonstrate that irinotecan was essential for Erbitux's effectiveness.

Around this time, Bristol-Myers Squibb entered into a $2 billion agreement with ImClone, agreeing to purchase 19.9 percent of ImClone's stock for $1 billion (an approximately 75 percent premium over the market price) and offering an additional $1 billion in milestone payments connected to Erbitux development and commercialization.

While the combination of Erbitux and irinotecan demonstrated a 22.5 percent response rate (measured by tumor shrinkage) and the single agent Erbitux study showed a 10.5 percent response rate, the results of the two trials were not statistically distinguishable. The interpretation of this finding is that ImClone had failed to demonstrate that combination therapy with irinotecan was necessary for Erbitux to be effective. Furthermore, missing data documenting that patients in the combination study were refractory to irinotecan led the FDA to question whether it was Erbitux or irinotecan that was responsible for tumor shrinkage in that study.

Without strong data demonstrating that combination therapy with irinotecan was necessary or that Erbitux was effective against tumors refractory to irinotecan, the FDA rejected Erbitux's first BLA. Erbitux was eventually approved after performing a new study and recovering and submitting much of the missing data.

a general rule, Phase IIa trials tend to address expansion and confirmation of data from Phase I trials—absorption, metabolism, and pharmacodynamics—whereas Phase IIb trials resemble small-scale Phase III trials in their evaluation of safety and clinical efficacy in large populations.

PHASE III

Phase III testing is the largest and most expensive clinical trial phase, and is intended to verify the effectiveness of a drug for the condition it targets, based on statistical end points established in Phase II trials. Phase III trials also continue to build the safety profile of drugs and record possible side effects and adverse reactions resulting from long-term use.

Phase III trials are tightly controlled, preferably double-blind, studies usually with at least 1,000 patients. In double-blind studies, neither patients nor the individuals treating them know whether the active drug or an alternative such as a placebo is being administered.

Relative to Phase I and Phase II trials, the larger and ideally more diverse populations used in Phase III trials are necessary to determine if certain types of patients develop side effects or do not respond to treatment.

Two successful Phase III trials are generally required to ensure the validity of the studies, although a single trial may suffice if the results are extremely strong. Phase III testing averages between three and four years.

APPROVAL

Assuming a drug reaches the desirable end point in Phase III trials, the sponsor company will then file a new drug application (NDA) or biologic license application (BLA), which contains detailed information supporting the efficacy and safety of the drug. NDAs are submitted to the Center for Drug Evaluation and Research (CDER) and describe small molecule therapeutics that can be discretely defined. BLAs covering therapeutic applications such as protein-based drugs, growth factors, and antibodies are also submitted to CDER. BLAs covering other purified biological products such as blood products, vaccines, gene therapy vectors, and antitoxins are submitted to the Center for Biologics Evaluation and Research. The decision of which FDA center processes an application depends on the availability of appropriate expertise to evaluate a drug.

Following NDA/BLA submission, a drug has a better than 70 percent chance of being approved. At the FDA, a review team of medical doctors, chemists, statisticians, microbiologists, pharmacologists and other experts evaluates whether submitted studies demonstrate that a drug is safe and effective for its proposed use.

An application must provide sufficient information, data, and analysis to permit FDA reviewers to determine if a drug is safe and effective for the proposed use(s), if the benefits of the drug outweigh the risks, that the proposed labeling is appropriate, and if the methods used in manufacturing and the controls used to maintain quality are adequate to preserve the drug's identity, strength, quality, and purity.

Approval of an application can take anywhere from two months

Drug Development Times

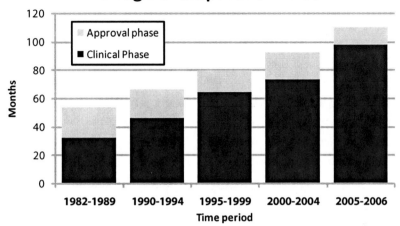

Figure 8.2 *Protein-based therapeutic development times*
Source: Grabowski, H. Follow-on biologics: data exclusivity and the balance between innovation and competition. *Nature Reviews Drug Discovery*, 2008. 7:479-488

to an extreme of several years, if the FDA requests additional information. The Hatch-Waxman Act permits day-for-day recovery of patent life for time spent waiting for FDA approval (see *Hatch-Waxman Act* later in this chapter).

Following FDA approval, a company may market and distribute a drug to the patient population determined in Phase III trials. At this point, a drug is likely protected by a patent that extends 20 years from the date of patent application, which was sometime before clinical trials began. Post-approval patent life-spans often range from 8-12 years.

Failure to receive approval usually means discussions with the FDA over trial design. A company can decide whether it is cost effective to run new trials and continue seeking approval. Alternatively, a company can shelve a drug, sell it to another company, or enter into an alliance with a partner and give away some part of future revenue in exchange for cash or assistance for further development.

PHASE IV

Once a drug is on the market, the sponsor must continue to perform observational studies in an ongoing evaluation of the drug's safety during routine use. Follow-up Phase III trials to confirm the safety and efficacy of drugs approved under accelerated development and review are sometimes referred to as Phase IV trials. In other cases, the safety and effectiveness of drugs may be monitored in applications other than those originally approved by the FDA.

Additional uses for many drugs are found after their initial launch. Accordingly, the objective of many companies is to find a good first indication for a drug in order to gain FDA approval. This demonstration that the drug has passed the FDA's safety assessment can be leveraged to gain marketing approval of additional indications for which a drug is effective.

An inherent risk of Phase IV trials is that they may reveal new data on side-effects or dangers associated with a drug. Failure to quickly and transparently respond to such information can provide a base for future tort lawsuits.

OFF-LABEL USE

While sponsoring companies are not allowed to advocate usage for non-FDA approved indications, physician-initiated "off-label" use can expand the market for a drug. For example, Cephalon's narcolepsy drug Provigil is largely used as a non-sedative treatment for depression.

A potential problem for popular drugs is that they may be prescribed to populations that are larger and more diverse than the clinical trial participants, and for conditions beyond the original indications. In some cases, the side effects that emerge can cause a drug to be shelved for all uses, underscoring the importance of monitoring off-label use. The Box *Fen-Phen: Risks of off-label use* illustrates how uncontrolled off-label use can lead to market withdrawal.

Cephalon's management of off-label use of its cancer painkiller Fentora is an example of proactive communication to control off-label use. After reports of four deaths related to the use of Fentora to treat conditions such as headache—not an approved indication for the drug—Cephalon sent letters to doctors reaffirming that Fentora was only to be used for cancer pain that is untreatable by regular doses of pain medicines. The Box *Avastin: Controlling off-label use* in Chapter 13 illustrates how off-label use can be attenuated to potentially encourage patients to use more expensive drug options.

Box

Vioxx: Anticipating and disclosing side effects

In September 2004 Merck voluntarily pulled Vioxx, a drug generating more than $2 billion in annual sales, from the market following evidence that the drug could be responsible for heart complications seen in long-term Vioxx users. In the following months thousands of lawsuits were filed against Merck, alleging that Merck failed to adequately warn patients and intentionally hid data on dangerous side effects for years.

When Vioxx and other so-called Cox-2 inhibitors were first discovered, they were hailed as next-generation anti-inflammatory drugs. Comparable first-generation anti-inflammatory drugs such as aspirin act by inhibiting both Cox-1 and Cox-2 enzymes. Inhibition of Cox-2 without affecting Cox-1 was hoped to reduce inflammation without the intestinal irritability associated with inhibiting both enzymes.

The mutually-opposing activities of Cox-1 and Cox-2 enzymes suggest the potential for vascular complications. Cox-1 and Cox-2 play counteracting roles in affecting narrowing of blood vessels and clot formation: inhibition of Cox-1 impedes blood vessel narrowing and clot formation, while inhibition of Cox-2 inhibits blood vessel dilation and enables clot formation.

Early trials in 2000 comparing Vioxx and naproxen, a non-specific Cox-1 and Cox-2 inhibitor, showed increased incidence of heart complications with Vioxx, but these were attributed to the possibility that naproxen offered greater protection against these complications. In 2002 the FDA requested a labeling change stating that these initial findings should be included on the label.

In a later trial investigating the ability of Vioxx to prevent the formation of colon polyps, the group receiving Vioxx experienced roughly twice the rate of serious cardiovascular side effects such as heart attacks and stroke as the untreated group. Unlike the aforementioned comparative study, this placebo-controlled study was able to discern the negative effects of Vioxx, illustrating the importance of using placebos in medical research. This finding led to the March 2004 withdrawal. Shortly thereafter, other manufacturers of Cox-2 inhibitors added warning labels to their drugs but declined to withdraw them.

Merck's legal exposure from Vioxx lawsuits threatened to bankrupt the company. The complementary box in Chapter 15, *Preventing a product recall from bankrupting a company*, describes the processes Merck used to limit their liability and reduce the impact of the recall.

> ### Box
> # Fen-Phen: Risks of off-label use
>
> Fen-Phen is the FDA-unapproved combination of two medications that were independently FDA approved for mutually exclusive use. The combination of fenfluramine, a drug approved for short-term treatment of obesity, and phentermine, a member of a class of drugs traditionally used for treatment of morbid obesity, that was intended to reduce obesity and counteract the sedative effects of fenfluramine.
>
> A series of scientific articles promoting the off-label, long-term use of Fen-Phen led to use by weight-loss clinics and prescription by physicians. The combination became so popular that fenfluramine use increased over 300 percent from 1995 to 1996, the largest single-year increase of any drug in the United States.
>
> In 1996, the Mayo Clinic reported the first diagnosis of Fen-Phen-induced heart valve damage. In the following months, dozens more cases were reported. In 1997 the FDA formally requested the voluntary withdrawal of fenfluramine and chemically-similar dexfenfluramine because of their implication in heart valve damage.
>
> The inappropriate use of fenfluramine caused, or at least hastened, its removal from the market. The heart valve symptoms associated with long-term use of Fen-Phen likely would not have occurred with short-term use of fenfluramine for its FDA-approved indication.[1]
>
> ---
>
> 1 Hutton, A.W., Dudley, L. A chronology of the Fen-Phen disaster and MDL 1203 litigation update. Hutton & Hutton LLP, 2001.

Another issue related to off-label use is the limited ability to prevent generic drugs from being sold for patented off-label applications. A company holding a patent for an FDA-approved use of a drug can, for example, potentially use the patent to block sales of generic drugs on the basis that the generic drug's labelling would advocate patent infringement. The same argument does not hold for a patent covering off-label uses, as the generic drug's label would not list applications which are not FDA-approved, and would therefore not advocate patent infringement.

ACCELERATED APPROVAL

Accelerated approval makes promising products for life threatening diseases available on the market prior to formal demonstration of patient benefit. Whereas the traditional approval process requires that clinical benefit be shown before approval can be granted, accelerated approval allows a new drug application to be approved before measures of effectiveness that would usually be required are available. Surrogate endpoints—indirect measures of effectiveness such as laboratory findings—are used to show the strong potential for effectiveness in accelerated development and review. An important element of accelerated development and review is that testing must continue after the drug is approved to demonstrate its projected safety and efficacy. Failure to meet projected endpoints can result in withdrawal from the market.

FAST TRACK

The fast track designation is often confused with accelerated approval, but the two vary greatly in the circumstances under which they are granted and the regulatory processes they impact.

Whereas accelerated approval makes experimental drugs available for life threatening diseases based on surrogate markers (indirect measures of efficacy such as tumor shrinkage), fast track is a process for interacting with the FDA during drug development. Fast track status grants drug developers scheduled development planning meetings with the FDA, the option of requesting the use of surrogate endpoints in evaluating studies rather than survival or other hard demonstrations of efficacy, and the option of submitting a BLA or NDA in sections on a rolling basis and authorizing the FDA to begin review of the application prior to its completion. Fast track status does not mean that a drug will receive faster approval, although priority review may be granted.

A recent study by the Tufts Center for the Study of Drug Development found that the fast track program reduced drug development time by nearly three years.[2] The study also noted that approval rates for drugs in the fast track program were roughly the same as

2 FDA's Fast Track Initiative Cut Total Drug Development Time by Three Years. November 13, 2003. http://csdd.tufts.edu/NewsEvents/RecentNews.asp?newsid=34.

Box

AIDS and accelerated approval

In the mid-1980s the FDA put forward a proposal to permit speedier access to experimental drugs in an effort to satisfy the needs of a growing population of AIDS patients frustrated by the complete lack of medications to treat HIV (the virus that causes AIDS) infection. The resulting Treatment Investigational New Drugs policy made promising new drugs available to desperately ill patients as early in the development process as possible.

After Phase I testing of the potential AIDS drug AZT in 33 people showed encouraging results, a Phase II trial commenced with 300 people given AZT or a placebo. Seven months into the trial 20 people had died; 19 who were given placebo and only one who was given AZT. This stunning demonstration of safety and efficacy compelled the FDA to permit a treatment protocol and AZT was administered to an additional 4,000 AIDS patients prior to its approval.

The second AIDS drug, ddI, was distributed to needy patients under the parallel track program. This program makes investigative drugs that have shown promise in preliminary trials available to patients whose condition prevents them from participating in clinical trials, while clinical trials are occurring. 16,000 people were given ddI for free through this program.

There were several problems with the parallel track program. The number of deaths associated with ddI, due to the advanced illness of people in the parallel track, rather than drug effects, nearly derailed the program. Additionally, because ddI was distributed directly from the producer but not through pharmacies, only patients of doctors with sufficient time and the ability to manage the necessary paperwork and follow-up had access to the drug. Furthermore, acute care facilities such as hospital emergency rooms did not have access to parallel track drugs.

Accelerated approval emerged to resolve issues in the parallel track program. Companies that met FDA requirements could gain marketing approval instead of simply gaining the right to give drugs away. The key provision for accelerated approval was that companies had to agree to perform certain clinical trials following accelerated approval.[1]

1 Delaney, M., Accelerated approval: Where are we now?, in *Research Initiative /
Treatment Action!*, 2002. pp. 18-22.

those without fast track designation, but those fast track drugs which failed to receive approval failed faster than non-fast track drugs. Companies also experienced a short-term (as short as one day) stock price increase following announcement of fast track designation for their drugs.[3]

PRIORITY REVIEW

The priority designation is intended for products addressing unmet medical needs. Reviews for new drug applications are designated as either standard or priority. A standard designation sets the target date for completing all aspects of a review and FDA approval decision at 10 months after the date the application was filed. A priority designation sets the target date for the FDA action at 6 months. Popular drugs which were granted priority review include Celebrex, Gleevec, and Taxol.

MARKET EXCLUSIVITY

In addition to market exclusivity derived from patents, the FDA also grants market exclusivity to drugs meeting special conditions. The motivation to grant exclusivity is to foster innovation and promote the development of drugs for applications that might otherwise offer insufficient motivation.

NEW DRUG PRODUCT

New drug product exclusivity (also known as new molecular entity protection), provided by the Federal Food, Drug, and Cosmetic Act, grants the holder of an approved NDA limited protection from new competition in the marketplace for the innovation represented by its approved drug product. A five-year period of exclusivity is granted to NDAs for new drug products—products containing chemical entities never previously FDA approved alone or in combination.

---

3 Cohen, F. The fast track effect. *Nature Reviews Drug Discovery,* 2004 3:293-294.

NEW CLINICAL INVESTIGATION

A 3-year period of exclusivity is granted for drug products containing previously approved active elements, when the application contains reports of new clinical investigations that were essential to approval of the application. For example, changes in an approved drug product that affect its active ingredient(s), strength, dosage form, route of administration or conditions of use may be granted exclusivity if clinical investigations were essential to approval of the application containing those changes.

PEDIATRIC USE

Pediatric use exclusivity is the only form of exclusivity that provides extensions which initiate at the termination of other exclusivity protection (e.g., new drug product or new clinical investigation) or patent protection. Sponsors that complete FDA-requested pediatric clinical investigations can be granted two separate six month extensions.

Companies do not need to gain, or even seek, new marketing approval to receive pediatric exclusivity extensions, they need only perform the requested studies. In 2007 AstraZeneca received a six month pediatric extension for its breast cancer drug Arimidex for investigating the therapeutic potential of Arimidex in pediatric conditions that result from increased estrogen production. AstraZeneca's studies reportedly did not show measurable benefits in treating the investigated conditions, leading them to forgo pursuing marketing approval for those indications.

OVER-THE-COUNTER

The political motivations to provide incentives for over-the-counter (OTC) drug sales without a prescription are to improve patient autonomy and accessibility to health care, and to reduce the cost of health care. These benefits must be weighed against hazards such as the potential for individuals to improperly self-diagnose their health conditions and needs, or the benefits of physician screening. Some conditions may be indicative of future health problems and while their direct treatment might not necessitate a physician visit, proactive screening for associated conditions can be beneficial.

Some of the criteria necessary to demonstrate safety and efficacy of an OTC drug are distinct from those applied to prescription drugs. The first consideration is whether or not patients can be expected to effectively diagnose themselves. The second requirement is that a patient must be able to read the product label and understand the indications and method of use of the drug. Third, it is necessary to determine if the OTC drug is safe and effective when used without supervision. Unlike prescription drugs, for which a doctor can monitor patient progress, OTC drugs are liable to be used in the absence of any supervision, raising the possibility of use not in accordance with labeling.

The motivations for drug companies to apply for OTC approval for a prescription drug—called an OTC-switch—are manifold. First, increased patient access can dramatically improve sales volume. Second, if the FDA deems that additional trials are necessary to prove safety and efficacy sufficient for OTC administration, applicant firms can be granted a 3-year exclusivity for the OTC market. Finally, drug companies may initiate an OTC-switch knowing that if they do not, a generic company may initiate a switch and obtain exclusivity.

ORPHAN DRUGS

Individuals affected by rare disorders often require many tests to confirm diagnosis and often must travel great distances to reach specialists with necessary expertise to diagnose and possibly treat the disease. Accordingly, drugs for so-called orphan diseases can have a significant impact on the quality of life of affected individuals and their families.

The Orphan Drug Act, enacted in 1983, provides incentives for companies that develop treatments for conditions affecting fewer than 200,000 Americans, or those that affect more than 200,000 Americans for which there is no reasonable expectation that the cost of research and development will be recovered from U.S. sales. Companies are given clinical testing grants, tax credits for the costs of clinical trials, and seven years of market exclusivity for approved products, regardless of patent protection. Similar programs exist in Japan, Australia, and the European Union. The effectiveness of the Orphan Drug Act has been significant. Whereas 10 orphan drugs

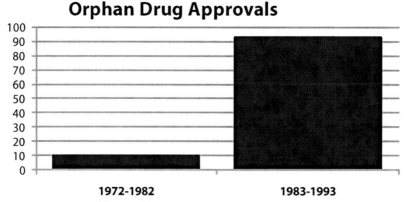

Figure 8.3 *Impact of the Orphan Drug Act*
Source: FDA

were approved in the decade preceding the Orphan Drug Act, 93 orphan drugs were approved in the decade following enactment (see Figure 8.3).

Developing an orphan drug can give a company the necessary expertise in drug development, clinical trials, and distribution that will be necessary for future development. A motivation for young biotechnology companies to develop orphan drugs is that they can develop expertise and technologies that can be used for numerous conditions, while working in a relatively supportive and noncompetitive environment. The relatively small number of physicians specializing in individual orphan diseases also makes it easier for a small sales force to effectively reach them, reducing sales and marketing expenses. According to the Tufts Center for the Study of Drug Development, biotechnology firms accounted for 65 percent of research spending on orphan drugs between 1998 and 2001.

Tackling a rare disease usually means that the number of unexpected side effects is less than for a broad-spectrum drug, simply because smaller populations have less variance than large ones. Furthermore, side effects may be more acceptable and a partial cure may also be accepted when compared to the severity of untreated disease. The income derived from orphan drug sales can also help fund future non-orphan drug development and the associated larger clinical trials. It is also possible to define a subset of a disease affecting a larger population as an orphan disease. A company can apply for orphan drug status on a medication for a specific and relatively uncommon type of cancer, for example, even if it has more general applications. Salami slicing is the technique of filing for orphan status for treatment of one rare disease, then adding new orphan designations for different applications of the same drug.

Additionally, despite being developed for small populations, not all orphan drugs serve small markets. In 2006, 19 orphan drugs

---

Box

Genzyme: Building an enterprise on orphans

Genzyme Corporation built a substantial portion of its enterprise by focusing on a therapeutic category which many other companies ignore: orphan drugs.

Genzyme's most lucrative niche is in lysosomal storage disorders. Cerezyme and Fabrazyme are enzyme-replacement therapies for Fabry disease, Aldurazyme is an enzyme-replacement therapy for Mucopolysaccharidosis I, and Myozyme is an enzyme-replacement therapy for Pompe disease. These treatments comprised nearly half of Genzyme's $3.8B revenues in 2007, with Cerezyme accounting for the largest share at $1.1B.

An issue facing Genzyme is that its drugs are very expensive. A year's dosage of these orphan drugs costs approximately $200,000. Careful management of drug pricing and access is crucial, as loss of insurance coverage, or loss of public and political support, could derail the company's business model.

Even with insurance coverage, the burden on patients can be significant. Genzyme defends its pricing by explaining that clinical trials for orphan drugs are very expensive, often requiring patients to be repeatedly flown to distant trial centers, and the drug products themselves are expensive to manufacture. Without a large population to generate sufficient revenues, the per-patient cost for orphan drugs must therefore be high. Genzyme also takes steps to ensure that no patient is denied access to their drugs. They offer numerous free drugs programs, covering patients in developing countries and those who are unable to access treatment despite living in countries with reimbursement programs.[1]

---

1 Free Drug Programs. Genzyme Corporation, 2008. http://www.genzyme.com/commitment/patients/free_programs.asp.

Box
The disproportionate returns of rare disease research

Beyond the humanitarian benefits of developing treatments for illnesses that would otherwise escape notice, an additional benefit of rare disease research is that rare diseases often yield scientific insights disproportionate to the small number of affected individuals. An example of the potential scientific and medical gains stemming from the study of rare diseases is the discovery of prions.

Investigating a single tribe in Papua New Guinea inflicted with a fatal mental illness that led to loss of coordination accompanied by dementia, Stanley Prusiner was able to trace the disease to the ritual consumption of deceased relative's brains. The protein factor responsible for the disease, similar to the agent causing mad cow disease, provided the first example of protein as genetic material. Prusiner was granted the 1997 Nobel Prize for Physiology or Medicine for his work on these infective proteins.

generated revenues in excess of $1 billion. The high revenues from orphan drugs can derive from expansion into additional markets (either through off-label use or by gaining approval for new indications) or the ability to demand a high price due to a drug's efficacy and lower cost compared to alternative treatments.

GENERIC DRUGS

A generic drug is a version of a pioneer, or brand-name, drug that is produced by a second party. When the patent on a pioneer drug expires, competitors are free to market generic versions, provided they can receive FDA approval. Based on historical trends, generic drugs can be expected to capture as much as 60 percent of pioneer market share within one year of market entry. Furthermore, generic prices average about 61 percent of innovator drug prices in the first month of entry and drop to 37 percent within two years.[4]

To encourage the development of generic drugs, the Hatch-Waxman Act provides 180 days of generic market exclusivity to the first company that can gain approval for a generic drug. This exclusivity

4 Grabowski, H.G., Vernon, J.M. Effective patent life in pharmaceuticals. *International Journal of Technology Management*, 2000. 19(1/2):98-120.

Figure 8.4 *Impact of generic entry on drug price*
Source: FDA analysis of IMS data from 1999 through 2004

commences from the date of initiation of commercial marketing to permit time to amass sufficient stock for distribution. A 75 day deadline to initiate marketing was added in 2003 to prevent first-approval generic firms from indefinitely blocking others by postponing marketing.

The approval process for generic drugs is simpler than for pioneer drugs. Generic drug applications may avoid the lengthy and costly clinical trial process and can receive approval if they can be proven to be bioequivalent to a clinically proven pioneer drug.

TRADITIONAL PHARMACEUTICAL DRUGS

For the purposes of determining bioequivalence there are two categories of drugs: small-molecule synthetic drugs and biologics (see *Biotechnology vs. Traditional Pharmaceutical Drug Development* in Chapter 4). Small-molecule synthetics are made using defined processes that result in a defined yield of product that is relatively pure. Biologics are derived from living sources and are much harder to objectively characterize. Examples of biologics include vaccines, gene therapy products, blood products, and tissues for transplantation. These preparations may be relatively unpure and the yield and precise nature of the product can also be difficult to define. The lack of consistency of biologic drugs does not imply that they are dangerous or ineffective; it just makes it more difficult to objectively compare pioneer and generic versions.

Generic manufacturers of small-molecule synthetic drugs must demonstrate identical chemical structure, purity, and concentration to the original drug. Generic drugs must also have the same pharmacokinetic profile as pioneer versions: the generic compound must be active at the same concentrations and for the same duration as the original compound. To gain regulatory approval for small-molecule synthetic drugs, generic manufacturers must file an abbreviated new drug application (ANDA) and use the same procedures, tools, and technologies as described in the original patents.

It can take two to five years to demonstrate bioequivalence and obtain approval for manufacturing facilities before an ANDA is granted. The production process and the source materials are strictly regulated. Just as two cooks following the same recipe may not produce identical products, generic manufacturers may face difficulty in producing a drug from the instructions in the original patent. In an interesting twist, generic producers may modify the production method and obtain patent protection for their improvements, potentially providing protection from other generics.

Biologic Drugs

There is no established process by which generic biologics can be demonstrated to be identical to pioneer drugs. The lack of a means for generic manufacturers to leverage clinical trial data from pioneers raises the cost of market entry and in many cases enables pioneers to avoid price competition long after patents expire. In 2005 Sandoz sued the FDA for their lack of progress in developing generic biologic approval guidelines, accusing the agency of relegating Omnitrope, their generic version of human growth hormone, to regulatory limbo. Omnitrope was eventually approved on May 31st 2006 as a "follow-on protein product," but not as a generic biologic.

The European Agency for the Evaluation of Medicinal Products, the European analog of the FDA, released approval guidelines for "biosimilars" in 2007. Recognizing that biosimilars are "similar but not identical" to their novel counterparts, the guidelines require that comparative studies demonstrate no meaningful differences between the safety and efficacy of the pioneer and biosimilar products.

A complicating issue for selling biosimilars in Europe is that in several countries pharmacists are not permitted to automatically substitute biosimilars. They must dispense the specific drug product prescribed by physicians. Patients, physicians, and pharmacists have no incentive to choose a biosimilar over a pioneer biologic, and

---

Box

Biologic manufacturing: The significance of small changes

Why are generic biologics so hard to regulate? The challenge stems from the size and complexity of these drugs. Figure 4.1 in Chapter 4 illustrates the size difference between pharmaceutical and biologic drugs. Amgen's Epogen, a biologic drug, is more than 500 times larger than aspirin, a pharmaceutical drug. This size difference alone makes it difficult to determine if a generic version of Epogen is identical to Amgen's original. Furthermore, protein-based drugs undergo subtle modifications—which can have profound implications on drug activity—that may be influenced by subtle changes in production methods and conditions. Additionally, many biologic drugs are complex mixtures. Subtle differences in production methods or ingredients can alter a biologic drug's safety and efficacy profile.

The case of Ortho Biotech's (a unit of Johnson & Johnson) Eprex illustrates the challenges of maintaining consistency in biologic manufacturing. Eprex is a version of epoetin alfa, similar to Amgen's Epogen, and is marketed outside the U.S. In 1998 the EMEA requested that the Eprex formulation be changed to eliminate the stabilizer human serum albumin, out of concern that it could transmit a variant of Creutzfeldt-Jacob disease (the human equivalent of mad-cow disease). Ortho Biotech responded by substituting Polysorbate 80 for human serum albumin.

As cases of immunogenetic responses, including a rare form of anemia called pure red cell aplasia, emerged among patients taking Eprex, Ortho Biotech initiated a five-year, $100 million investigation, which ultimately discovered that Polysorbate 80, the human serum albumin substitute, was reacting with uncoated rubber stoppers used in pre-filled syringes (one of the dosage types), and was leaching allergenic organic compounds from the rubber stoppers.

the only party with an interest in using biosimilars is the healthcare payer. Therefore, while biosimilar producers can realize reduced development and approval expenses, their marketing expenses are likely to be comparable to, or greater than, those of any other novel drug product.

The general model of generic biologic approval in the United States will likely be as "follow-on" drugs requiring limited clinical trials to demonstrate similarity to pioneer biologics. The impact of this system is that individual generic biologic drugs are likely to differ slightly from pioneer drugs and each other, potentially offering therapeutic improvements, opportunities for personalized medicine, and ultimately complicating prescription decisions.

In addition to uncertainty over the mechanisms of generic biologic approval and marketing, the expected cost savings are also in dispute. It is anticipated that generic biologics will have higher development costs, higher approval costs, and higher manufacturing costs than generic pharmaceutical drugs. If generic biologics have different efficacy and safety profiles than pioneer biologics, then they may end up with a smaller relative market share than generic pharmaceuticals are able to capture. The impact of these potential increased costs and decreased profits is that consumers will not realize the same savings with generic biologics that they have seen with generic pharmaceuticals. It is estimated that generic biologics will be sold at a 20 to 30 percent discount, versus generic pharmaceuticals which may be sold for discounts as high as 90 percent.

IMPEDING GENERICS

A strategy to impede generic competitors is to trigger a 30-month stay of generic approval. Patents that cover a drug's production or its use may be submitted to the FDA's *Approved Drug Products with Therapeutic Equivalence Evaluations*, or *Orange Book*. Because generic manufacturers must certify that no patents are infringed in their applications, generic versions of pioneer drugs cannot receive FDA approval until all applicable *Orange Book* patents have expired or been invalidated.

A generic company wishing to market their drug prior to expiration of all listed *Orange Book* patents must notify the original NDA sponsor of the bases for invalidity or non-infringement. The pioneer firm then has 45 days to sue for infringement, which may trigger a 30-month stay of generic approval. In the event of an infringement claim from the pioneer, the generic manufacturer, potentially having produced sufficient product for distribution, must choose to wait for the contested patent to expire, or may challenge the patent and risk an automatic 30-month delay during litigation.

In response to repetitive and sometimes frivolous *Orange Book* patent listings, in August 2003 the FDA instituted new rules on patent listing requirements and the 30-month stay, limiting it to no more than one 30-month stay per drug and defining the types of patent claims that cannot, and must, be listed.

Pioneer drug developers are also able to frustrate generic competitors by exploiting a loophole in federal law. The first firm to gain approval for a generic version of a drug may be granted 180 days of generic marketing exclusivity. Pioneer firms can reduce the impact of this marketing exclusivity by licensing a third party to sell the pioneer product, rebranded and marketed under a generic name, limiting generic profits during the generic marketing exclusivity period. Because these so-called "authorized generics" are simply relabeled versions of the pioneer drug, they are not subject to additional FDA approval, so the FDA cannot block their sale. It is estimated that one third of branded-drug firms had launched authorized generics between 2005 and 2007, and that number is expected to grow to 44 percent between 2008 and 2010.[5]

Hatch-Waxman Act

The Hatch-Waxman Act, enacted in 1984, promotes innovation by fostering competition and allowing patent owners to recover time spent in clinical trials and FDA regulatory review. Measures easing the development of generic alternatives to pioneer drugs are balanced by patent term extensions that support pioneer drug manufacturers. The impact of the Hatch-Waxman Act has been significant. According to the Generic Pharmaceutical Association, the generic share of the prescription drug market has grown from 19 percent in 1984 to 53 percent in 2003, but accounts for only 12 percent of pharmaceuti-

5 Counter Generics Strategy, Planning, and Execution. *Cutting Edge Information*, 2008.

cal costs.

Hatch-Waxman permits generic manufacturers to cite safety and effectiveness data from pioneer FDA applications, relieving the burden of performing lengthy and expensive clinical trials. To use pioneer safety and efficacy data, a generic drug must be demonstrated to be bioequivalent to a pioneer drug. In an exception to patent law, generic manufacturers are permitted to initiate clinical tests (necessary to demonstrate bioequivalence) before pioneer patents expire. As an incentive to promote the development of generic drugs, the first company to gain approval for the generic form of a drug is granted 180 days marketing exclusivity before other generic firms may enter the market.

To balance the benefits given to generic manufacturers, Hatch-Waxman also includes considerations for pioneers. A unique aspect of drugs, relative to most other marketable products, is that drugs cannot be marketed until they are proven safe and effective. This restriction places a significant financial and time burden on drug development firms, with a greater impact on innovative drugs requiring extensive clinical trials and possibly protracted FDA review. Hatch-Waxman grants an extra half-day restoration of patent life for every day of clinical trials and day-for-day restoration for the FDA review period. There are two restrictions on restoration: The effective life of a drug patent cannot exceed 14 years, and the total time restored cannot exceed five years.

If application of Hatch-Waxman extensions is anticipated, patenting and marketing plans must consider several limitations: a patent term can be extended only once, only one patent can be extended for a given regulatory review period, FDA authorization must be for the first commercial marketing use of the drug product's active ingredient, and extensions are limited to what is covered by the original FDA authorization.

While consolidation of multiple drugs under one patent may save money and simplify patenting, should more than one drug covered by a patent receive approval, only one will be eligible for extension. Claiming a single drug under multiple patents enhances patent protection but results in Hatch-Waxman extensions being limited to claims from only one patent. Contemplation of the potential value of numerous markets for a single drug is therefore important, because only the first marketed application is eligible for extension.

DRUG NAMES

Drugs have several types of names, each of which is used for a different context (see Table 8.3). Chemical and biological names are respectively used to describe the composition of small-molecule or biological drugs. Generic names are shared between branded and generic drugs to indicate common ingredients. The trade name of a drug is the proprietary name used by different firms to brand their products.

The chemical or biological name of a drug is determined using several conventions, the objectives of which are to provide scientific descriptions of the composition of a drug. The chemical name for ibuprofen, for example, is *2-[4-(2-methylpropyl)phenyl]propanoic acid*; the biological name for Amgen's Epogen is *recombinant erythropoietin*. The generic name of a drug is created using a specific nomenclature system and is used to identify generic versions of a branded drug. The generic name for Amgen's Epogen is *epoetin alfa*. Ibuprofen is the generic name for a drug which has been marketed under Advil (Wyeth), Motrin (McNeil), and other trade names.

Trade names are used to uniquely brand an individual company's version of a drug. A drug's proprietary or trade name must be

Table 8.3 *Drug names*

Chemical / Biological	Scientific description of drug compound
Generic	Simplified name based on drug function
Trade name	Branded name, protected by trademark law

Table 8.4 *Selected generic drug naming conventions*

Name element	Category	Examples
-vir	antivirals	acyclovir, combivir
-mab	monoclonal antibodies	cetuximab (ImClone's Erbitux), rituximab (Biogen Idec's Rituxan)
-rsen	antisense oligonucleotides	fomivirsen (Isis Pharmaceutical's Vitravene)

approved by the FDA and cannot imply efficacy. Trade names are protected by trademark law, preventing generic companies from using them even after patents expire, and encouraging pioneers to develop strong brand identity to extend their market dominance past patent expiration. Despite the restriction that trade names cannot imply efficacy, drug makers often select names with connotations aligned with a drug's intended use. Vick's Dayquil and Nyquil respectively suggest daytime or night time tranquility in treating cold and flu symptoms.

Other useful names such as abbreviations (EPO is a common substitute for epoetin alfa) may be used when appropriate. Drugs in development are referred to by their method of action (e.g., ACE inhibitor) or internal code name (e.g., MEDI-493 was an internal code name for MedImmune's Synagis).

DEPARTMENT OF AGRICULTURE

The USDA regulates plant pests, plants, and veterinary biologics. Because of the potential for plants to interact unfavorably with the environment, it is important to determine that a plant is safe to grow outdoors in order to prevent the release of plants that may harm other plants and animals or spread like a weed.

Several USDA agencies are involved in regulating and monitoring the use of biotechnology for agriculture. The Agricultural Marketing Service is responsible for administering plant variety and seed laws. AMS also offers laboratory testing services for genetically engineered food and fiber products. The Agricultural Research Service (ARS) is USDA's in-house science agency. ARS works to improve the quality, safety, and competitiveness of U.S. agriculture. The agency's biotechnology research includes introducing new traits and improving existing traits in livestock, crops, and microorganisms; safeguarding the environment; and assessing and enhancing the safety of biotechnology products. ARS also develops and provides access to agricultural resources and genomic information.

The Food Safety Inspection Service has responsibility for the safe use of engineered domestic livestock, poultry, and products derived from them. The Animal and Plant Health Inspection Service

(APHIS) is responsible for protecting agriculture from pests and diseases. APHIS regulates the movement, importation, and field testing of genetically engineered organisms through permits and notification procedures. In addition, APHIS Veterinary Biologics inspects veterinary biologic production establishments and licenses genetically engineered products. Another unit of APHIS, the Biotechnology Regulatory Services, monitors biotechnology industry trends and forecasts scientific advancements to better help regulate the industry.

The Foreign Agriculture Service supports the overseas acceptance of biotechnology and crops that have been reviewed by the U.S. government agencies to support U.S. farm exports and promote global food security. The Economic Research Service conducts research on the economic impact of genetically engineered organisms. The Cooperative State Research, Education, and Extension Service administers the biotechnology risk assessment program as well as research programs in gene mapping, sequencing, and biotechnology applications.

ENVIRONMENTAL PROTECTION AGENCY

The EPA regulates microbial and plant pesticides of chemical and biological origin, new uses of existing pesticides, and novel organisms that may have industrial uses in the environment.

Under the Federal Insecticide, Fungicide, and Rodenticide Act, EPA regulates the domestic manufacture, sale, and use of all pesticides, including those derived through modern biotechnology. Section 408 of the Federal Food, Drug, and Cosmetic Act gives the EPA the mandate to ensure that any pesticide residue in or on a food product falls within a safe limit. Either a safe limit (known as a food tolerance) is set, or an exemption from the requirement of a tolerance is granted. The Toxic Substances Control Act Biotechnology Program tracks all chemical substances or mixtures of chemical substances produced or imported into the United States. Genetically engineered microorganisms meet the program's legal definition of a "mixture" of chemical substances and are therefore regulated under this act.

Box

Starlink corn: Controlling biotechnology crops

In 1995 Plant Genetics Systems, later purchased by Aventis, developed a genetically modified corn variety containing a bacterial toxin to poison the corn borer, a pest that costs American farmers over $1 billion a year in damage and pesticide costs. The toxin, derived from a bacterium that naturally repels corn borers, was produced by the introduction of a gene called Starlink into the corn plants.

Starlink corn was subject to oversight by the USDA, EPA, and FDA. When EPA scientists evaluated Starlink corn, they noticed that the corn borer toxin in Starlink corn did not break down in the stomach as fast as analogs in other genetically modified plants. The Starlink toxin was also more resistant to cooking and food processing than other varieties. This meant that the Starlink toxin might survive food preparation and stomach acids long enough to enter the bloodstream, where it could potentially trigger an allergic reaction (it is worth noting that no cases of allergic reactions to Starlink have been noted).

Although the Starlink toxin did not resemble any known human allergens, the EPA nevertheless wanted the safety of Starlink corn in humans to be established prior to any use. The developers countered by invoking the EPA split registration rule that allows some pesticides to be used on animal feed but not human feed.

Setting restrictions in an attempt to control the potential spread of Starlink corn, the EPA required that Starlink corn be grown only for animal feed or nonfood use. Farmers were required to leave 660-foot buffer strips around Starlink corn fields to prevent the spread of pollen into neighboring fields and had to notify grain elevators prior to delivering Starlink corn. All farmers buying the corn had to sign agreements acknowledging that they understood the restrictions and would comply with them. Aventis agreed to conduct a survey of Starlink corn farmers to determine compliance and agreed to accept full liability in case anything went wrong.

Failures at all levels of control resulted in Starlink corn being detected in a range of consumer products such as tacos, corn chips, and muffin mix. Recalls of Starlink corn-containing products cost Aventis hundreds of millions of dollars. While there were no reports of anyone becoming seriously ill from eating Starlink corn, the incident revealed shortcomings in the oversight of genetically modified agriculture products. The EPA has since revised its guidelines to not accept any genetically modified food for use as animal feed unless it is safe for humans as well.

SECURITIES AND EXCHANGE COMMISSION

The mission of the Securities and Exchange Commission (SEC) is to protect investors, maintain fair, orderly, and efficient markets, and facilitate capital formation. The SEC's challenge in regulating biotechnology companies is that many biotechnology company disclosures are protected by the FDA, and are therefore not public information. Because of the sensitivity of information that the FDA is privy to, the FDA is limited in what they can disclose to the public. The SEC is responsible for ensuring that company public disclosures are accurate and complete, but the only way they can become aware of inconsistencies is if the FDA releases information to them. The SEC does not require companies to provide them with all the information they receive from the FDA.

The result of shielding drug developer-FDA communications is that while the FDA has access to the information necessary to enable investors to make informed decisions, it is not permitted to disclose this information. The SEC, which is responsible for policing public disclosures, often does not have sufficient information to judge the completeness or accuracy of disclosures. A company can, for example, release optimistic press releases despite receiving serious challenges of their study design from the FDA, without the FDA being able to make investors aware of the issues or the SEC being aware of an incongruity between public company statements and internal knowledge of the likelihood of drug approval.

Following a rash of misrepresentations by biotechnology companies, the FDA improved communications with the SEC and began releasing carefully worded press releases to inform the public of inconsistencies. An example of this cooperation was seen on March 14th 2003 when the FDA issued a public warning about SuperGen, Inc., stating that the company "exaggerates the efficacy of Mitozytrex and fails to include the significant risks associated with the use of the drug" and that "[n]o data submitted by the company provided evidence that Mitozytrex is superior to existing marketed formulations of [competing drug]."

Chapter 9

Politics

> One of the penalties for refusing to participate in politics is that
> you end up being governed by your inferiors.
> *Plato*

Political decisions can have a profound impact on biotechnology innovation. The biotechnology industry faces regulatory oversight in most of its operational activities, spanning from licensing research facilities and setting allowances for the use of radioactive materials or animal testing, through pre-and post-launch safety assessments (and, in the case of drugs, efficacy assessments). The regulatory rules influencing these activities ultimately stem from political decisions. In many countries the price for which drugs are sold is also influenced by government-backed organizations.

Governments around the world play crucial roles in influencing the development of their domestic biotechnology industries. Funding, tax breaks, and incentive plans are used to create beneficial market distortions and influence research and development. The war on cancer, for example, promoted cancer research by granting hundreds of millions of dollars in support. Incentives can also promote the development of drugs that would otherwise not have sufficient financial merit, the Orphan Drug Act (see *Orphan Drugs* in Chapter 8) provides incentives for companies to develop treatments for conditions for which there is no expectation that companies can otherwise recoup their R&D expenses. In a bold initiative to spur venture capital activity, countries such as France, Germany, Austria, and several U.S. states, have embarked on equity guarantee plans, partially guaranteeing qualifying venture capital investments against losses.[1]

Governments also play an important role in assuring the public that their regulatory protections are effective. The emergence of European resistance to genetically modified crops is often tied to the emergence of "Mad Cow" disease, or bovine spongiform encephalopathy (BSE), following relaxed British rendering practices in the 1980s that fed inadequately processed animal remains to herds of the same animal. Poor handling of the spread of BSE and reluctance to admit a link to human disease led to a mistrust of governmental regulation of food products. A recent Italian study on public perception bears out this sentiment: 84 percent of survey respondents favored continued research on medical biotechnologies whereas only 57 percent supported continued research on food biotechnologies. These opposing views are explained by a "perceived absence of adequate and publicly accountable procedures for the governance of innovation."[2]

At a national level, political decisions can create or eliminate markets through the regulatory and intellectual property protection mechanisms described earlier in previous chapters, defining the scope of the biotechnology industry. Bans on applications such as genetically modified crops and human cloning effectively eliminate them from the market. Alternatively, countries that do not protect patents display a noticeable absence of innovative biotechnology companies, but may have a strong generic drug industry.

SUPPORTING BIOTECHNOLOGY INNOVATION

Why is the U.S. the world leader in biotechnology? One significant factor is strong political support; the United States government contributes more funding to biotechnology than any other country, nearly exceeding the sum of the other G-7 nations. Intellectual property protection is also an important supportive element. The strong intellectual property policy credited with protecting innovation is set by Congress and administered by the U.S. Patent and Trademark

1 For more details, see: Raising EU R&D intensity. 2003, *Office for Official Publications of the European Communities*, Luxembourg.

2 Bucchi, M., Neresini, F. Why are people hostile to biotechnologies? *Science*, 2004. 304(5678):1749.

Box

State autonomy and federal laws

The political system in the United States gives individual states a great deal of autonomy enabling them to resist implementation of federal laws and enact strong regional laws. A prominent example of a state act that resists federal laws and provides regional incentives is California's Proposition 71, the California Stem Cell Research and Cures Act. This act affirms the legality of stem cell research and provides generous grant support for stem cell research.

In 2001 President Bush signed a bill significantly limiting the use of federal funding for stem cell research. Many scientists and non-scientists disagreed with this action, hoping to reap the significant medical advances promised by stem cell research. Shortly afterward the California legislature approved a bill that made stem cell research legal in California, affirming the legality of stem cell research. Proposition 71 was later introduced in a bid to support stem cell research for new cures and treatments and to prevent the exodus of scientists from California, home to the largest biotechnology industry in the union.

The proposition, passed in November 2004, calls for the establishment of the California Institute for Regenerative Medicine (CIRM), establishes the constitutional right to conduct stem cell research, prohibits the Institute's funding of human reproductive cloning research, and authorizes a bond issue to finance Institute activities up to $3 billion subject to an annual limit of $350 million. After passage of proposition 71, CIRM was almost immediately mired in litigation to prevent it from funding research, but the proposition set a strong precedent, motivating several other states to propose their own stem cell research funding programs.

Office. Furthermore, the alignment of incentives—funding basic research in academia and encouraging the transfer of technologies with commercial merit to industry—enables the U.S. to realize greater productivity per dollar spent on research than other countries (For additional discussion of the factors which facilitated the genesis and growth of biotechnology in the U.S., see Box *Why did the biotechnol-*

ogy industry start, and prosper, in the United States? in Chapter 16).[3]

BAYH-DOLE ACT

The Bayh-Dole Act, enacted in 1980, was cited by *The Economist* as one of the most significant pieces of legislation of the latter 20th century. Bayh-Dole reversed the presumption of title, specifically granting universities the right to license exclusive rights to patents emerging from federally funded research. Prior to Bayh-Dole the government was reluctant to grant ownership of federally funded inventions to inventing organizations, who would have been able to offer exclusive licenses to third parties, electing instead to offer non-exclusive licenses to interested parties.

In 1980 the U.S. government held title to 28,000 patents, but fewer than 5 percent of these were licensed to industry, presumably because companies were unwilling to invest in the research and development necessary for commercialization without exclusive rights to the technologies. The Bayh-Dole and Stevenson-Wydler Technology Innovation Acts were introduced, following the consensus that the public would benefit from a policy that permitted universities and small businesses to take ownership of inventions produced under federal funding and become directly involved in their commercialization. These acts allowed government contractors, small businesses, and nonprofit organizations to retain certain patent rights to government-sponsored research and permitted technology transfer to third parties.

The stated intent of the Bayh-Dole Act was to ensure that patents resulting from federally funded research would be broadly and rapidly available for all scientific investigation. The Bayh-Dole Act permits individual researchers to patent inventions and grant licenses for research supported by federal funds, creating an incentive for researchers and universities to ensure that their intellectual property is properly developed. Proceeds from these licenses must be used for scientific research or education. The emphasis on putting the results of government-sponsored research in the public domain promotes

3 Lawlor, M. Biotechnology and government funding: economic motivation and policy models Prepared for Science and Cents: The Economics of Biotechnology. *Federal Reserve Bank of Dallas*, April 19, 2002.

a pro-patent stance and reinforces the need for exclusive licenses to motivate industry to invest in the development of products for commercialization.

An important clause associated with Bayh-Dole is the govern-

Box

Directing innovation: The war on cancer

Cancer is the second leading cause of death in the United States, after heart disease and the leading cause of death among children between the ages of 1 and 14. Cancer-related expenses account for 10 percent of the total amount spent on disease treatment in the United States, and the disease is poised to overtake heart disease as the leading cause of death.

Tackling cancer is a difficult task, as the disease can be caused by a vast number of different conditions which result in the general category of uncontrolled cell growth. In his January 1971 State of the Union address President Nixon, recognizing the significant social and economic impact of cancer, made a special request for an additional $100 million to be added to the National Cancer Institute budget for cancer research. In October 1971 he converted the Army's Fort Detrick, Maryland, biological warfare facility to a cancer research center. Finally, the National Cancer Act was signed into law on December 23rd 1971, directing the National Cancer Institute Director to expand and develop a coordinated cancer research program and authorizing the first cancer centers.

While cancer as a disease category has not yet been cured, the impact of the initiatives has been significant. The death rate from cancer has been decreasing since the 1990s and several types of cancers have been effectively cured.[1] Whereas a diagnosis of cancer once meant impending death, today nearly half of all cancer patients can expect to live for five or more years after the diagnosis of cancer. Additionally, there have been substantial technology and knowledge spillovers into other disciplines. In one example, Leukemia drug Gleevec has been shown to have activity in diabetes.[2]

1 Sporn, M.B., The war on cancer: a review. *Annals of the New York Academy of Sciences*, 1997. 833(1):137-146.

2 Lassila, M. *et al.* Imatinib attenuates diabetes-associated atherosclerosis. *Arteriosclerosis, Thrombosis, and Vascular Biology*, 2004. 24:935.

ment's "march-in" rights, enabling the government to request (and, upon refusal potentially require) an owner or licensee of a patent developed using federal funding to issue a license to another party.

Box

Presidential politics and biotechnology

We have a profound responsibility to ensure that the life-saving benefits of any cutting-edge research are available to all human beings. Today we take a major step in that direction by pledging to lead a global effort to make the raw data from DNA sequencing available to scientists everywhere to benefit people everywhere. I urge all of the nations, scientists and corporations to adopt this policy and honor its spirit. We must ensure that the profits of human genome research are measured not in dollars, but in the betterment of human life.
U.S. President Bill Clinton, March 14th 2000

On March 14th 2000, with the completion of the human genome project in sight, U.S. President Bill Clinton and UK Prime Minister Tony Blair issued a statement advocating that raw data from the human genome project should be freely available. Investors in the United States interpreted this statement to indicate a stance against drug and other biotechnology patents and exited the sector, causing the NASDAQ biotechnology index to fall by nearly 14 percent. Biotechnology shares continued to fall for many months, until President Clinton reaffirmed the patentability of biotechnology inventions stating, "General information ought to be in the public domain … If someone discovers something that has a specific commercial application, they ought to be able to get a patent on it."

More recently, President George W. Bush issued a prohibition on the use of federal funding for research on all but existing stem cell lines. While this ruling does not prevent the use of state or private funding for stem cell research, it has negatively impacted investment and research in the field. It remains possible for privately funded companies to develop medicines based on stem cells, but since approval of those medicines lies with the FDA, a federal agency, the possibility exists that federal regulatory decisions will echo federal funding sentiments and restrict stem cell use.

March-in may be initiated due to a patent holder or licensee's slow progress in taking appropriate steps to apply a technology for health, safety, or other public needs. The NIH has received several requests to march-in on patents but refused to intervene, instead ruling that the cases were motivated by commercial and drug price disputes. No federal agency has exercised march-in rights to date, but it remains a potential risk for any company using federal funds to develop patents or licensing patents from federally funded researchers.

The Stevenson-Wydler Act, which was also enacted in 1980 and complements the Bayh-Dole Act, established the policy that government agencies should ensure full use of the results of federal investment in research and development, and requires federal laboratories to take an active role in the transfer of federally-owned and originated technologies to state and local governments and private firms. An amendment to Stevenson-Wydler, the Federal Technology Transfer Act of 1986, authorizes direct collaboration between government laboratories and private industry, an action that would have previously been considered a conflict of interest.

BALANCING INNOVATION INCENTIVES WITH ECONOMIC CONSTRAINTS

The debate over price controls and intellectual property protection is contentious and there are conflicting opinions on the impact of legislation in these areas. It is also worth noting that because price controls, patent laws, regulatory requirements, and biotechnology product import bans impact global trade, they may also be used as trade barriers to bolster domestic industries, under the guise of humanitarian or safety concerns.

Some argue that price controls enable countries to set manageable compensation rates for drugs, whereas others argue that setting prices too low can provide a disincentive for development of locally-endemic conditions, preventing the treatment of illnesses that price controls are supposed to facilitate. The case against strong intellectual property protection argues that weak intellectual property protection enables a strong foundational generic industry to take root. Opponents claim that strong intellectual property protection can favor foreign investment and help develop a robust innovative industry.

Prices and patent protection also direct innovation. Companies seeking profitable returns from their R&D investments target markets which can deliver sufficient revenues. If a drug treats a condition which is endemic in countries where price controls or poor intellectual property diminish the opportunity for profitable sales, a company may prefer to shelve the drug or potentially donate it to a non-profit organization for further development (See Box *Non-profit drug development* in Chapter 12).

PRICE CONTROLS

Price controls enable governments to set the prices at which they will reimburse companies for drugs. This can help manage healthcare budgets, but when there are disputes over compensation a drug company may simply opt to not sell a drug in a given market. As an example, in June 2006 Bristol-Myers Squibb Canada announced its refusal to market its colorectal cancer drug Erbitux in Canada due to an inability to agree on a price with the Patented Medicine Prices Review Board, which regulates the price of patented medicines. Genzyme is able to partially avoid price control issues by offering drugs for free in developing countries (see Box *Genzyme: Building an enterprise on orphans* in Chapter 8).

Price controls are also implicated in the observation that drug companies are increasingly submitting their drugs for approval in the United States prior to the European Union, and are focusing their R&D efforts in the U.S. as well. According to the Tufts Center for the Study of Drug Discovery, the relative increase in U.S. R&D investment stems from the lack of price controls.[4]

INTELLECTUAL PROPERTY

The World Trade Organization provides provisions for countries to issue a "compulsory license" to produce generic versions of patented drugs, in the event of a health crisis. Drug companies have generally resisted the implementation of compulsory licenses, fear-

---

4 Mitchell, P. Price Controls Seen as key to Europe's Innovation Lag. *Nature Reviews Drug Discovery*, 2007. 6:257.

ing loss of international markets and the potential that manufacturers will leverage compulsory licensing to produce more drug than is needed domestically and sell it to other nations.

Following several instances of anthrax-laden packages being sent in the mail in 2001, the United States government sought to assemble a stockpile of Bayer's Cipro to potentially treat 10 million exposed people. The government investigated issuing a compulsory license to enable it to purchase drugs from Indian generic firms as a bid to speed stockpile assembly, and to pay a lower price for the drug. The threat of compulsory licensing compelled Bayer to dramatically increase production and offer a lower price for the drug.

In a more extreme case, Abbott declared in March 2007 that it would not launch any more drugs in Thailand in response to the government's decision to not honor the company's patent for a new AIDS drugs and a heart drug. Facing a public relations backlash, the company reversed its decision in April, and eventually agreed to offer the AIDS drug below the price of generic versions, if Thailand agreed to honor their patents.

Another intellectual property concern is the manufacture of drugs in countries where they lack patent protection and subsequent export to more valuable markets. As described in the Box *Novartis cancels Indian investments over patent dispute* in Chapter 16, Novartis was unable to obtain an Indian patent for its drug Glivec (sold as Gleevec in the U.S.), which is widely patented in other countries. Interestingly, Novartis provides Glivec free to most patients in India. Because Indian manufacturers would be unable to compete with Novartis' free domestic distribution, their target markets would likely be in other countries, where they could potentially erode Novartis' market.

IV

The Business of Biotechnology

This section describes the commercial considerations of biotechnology companies, with the goal of enabling their effective management. The long development timelines and large up front capital investments required by biotechnology firms create unique challenges. Companies need to raise large sums of money—in several rounds—long before they are able to generate positive revenue streams.

Biotechnology companies must find a balance of resource allocation that combines a strong scientific base, sufficient financing, and relevant business expertise.

Success in biotechnology absolutely requires three elements: strong management, financing, and technology. Management is arguably the most important ingredient because of the vital need for proactive marketing, to maintain financial health, and to redirect or replace research efforts when necessary. In the absence of good management, otherwise sufficient financing and technology cannot realize their potential.

Beyond simply obtaining financing and other resources, and managing expenditures, management is also responsible for aligning research efforts with market needs. A common reason for commercial failure is development of products, which are often based on innovative science, for which profit-enabling markets do not exist (see Boxes *Biotechnology myth: Build it and they will come* in Chapter 13 and *Flavr Savr tomatoes: Operating in unfamiliar markets* in Chapter 15).

Because most biotechnology applications are based on innovation rather than price competition or modest improvements of existing applications, it is often necessary to invest heavily in research and development long before any revenues are generated. Intellectual property protection is therefore very important in biotechnology, securing the potential to achieve profitable revenues.

The knowledge-based and research-intensive nature of biotechnology companies gives them unique characteristics. The speed of technological development and the fundamental role of people and ideas make biotechnology companies harder to assess as investment opportunities. Startups require growth capital at an early stage because retained profits, if they exist, are usually insufficient to support the characteristically long development times. Additionally, as a company matures the management team will require different sets of skills to satisfy changing needs; early stage companies require multifaceted individuals to drive research and build relationships, whereas more mature companies need specialists who can manage relationships and support the development and commercialization of research leads.

Chapter 10

Biotechnology Company Fundamentals

I like to buy a company any fool can run because eventually one will.
Peter Lynch, Fidelity Magellan Fund Manager

T he unique conditions faced by biotechnology companies—the necessity for funding well in advance of revenues, need for strong intellectual property protection, and regulatory clearance requirements—influence individual companies and intercompany industry relationships. This chapter profiles some common characteristics in the structure of biotechnology companies to establish a framework for the later discussion of operational elements.

COMPANY FORMATION

There are several ways that new biotechnology companies form. Two common themes are a scientist in an academic or industrial position forming a new company to commercialize a novel idea, or an established company or university forming a new company to develop a potentially lucrative technology. In other cases, a company may be formed by entrepreneurs starting companies to pursue lucrative opportunities such as developing validated drug leads (see *Specialty Pharmaceutical / NRDO Models* in Chapter 12), finding new uses for existing drugs, or exploiting new scientific advances.

START-UP

Independent start-ups emerge from the union of a scientist or group of scientists with a commercial vision, and one or more financiers. The prototypical start-up involves a group of scientists with proven expertise who seek to commercialize a new technology. While a small amount of initial funding may be provided by "friends and family," the bulk of early capital comes from professional investors such as venture capitalists and angel investors. Professional investors also contribute valuable business expertise, guidance and networking in addition to the money they provide in exchange for equity. Because their ultimate goal is to gain a return on their investment— a goal which may be at odds with the founder's primary interests—it is necessary to align the goals of investors and founders. This topic is discussed in further detail in Chapter 11.

Genentech, founded in 1976 by venture capitalist Robert Swanson and biochemist Herbert Boyer, is an example of a biotechnology company that was initiated as an independent startup. In 1973 Boyer and fellow biochemist Stanley Cohen demonstrated the first expression of spliced genes in bacteria. Swanson approached Boyer with the proposal to form Genentech to commercialize this revolutionary technology. After failing to secure funding from the National Institutes of Health, Genentech turned to venture capital firm Kleiner and Perkins.

Staging their investments over milestones, Genentech's first goal was to validate the potential of gene splicing by demonstrating the ability to produce a human protein in bacteria. Following proof-of-principle production of the brain hormone somatostatin, Genentech later produced recombinant human insulin, the first marketed biotechnology product (see Box *Genentech: Commercializing a new technology* in Chapter 2).

Genentech wasn't able to reap the full rewards of its first innovative product. In order to obtain the financing necessary to develop human insulin, Genentech had licensed manufacturing and distribution rights to Eli Lilly. Three years later, Genentech became the first biotechnology company to independently manufacture and market a product as they leveraged their proven developmental expertise and a better financial position to introduce Protropin (human growth hormone). Today Genentech is one of the largest biotechnology companies.

Technology Transfer

In academic settings technology transfer refers to the transfer of research findings from academic laboratories to the commercial marketplace for public benefit.

The rationale for technology transfer is that the various institutions involved in biotechnology product development are stratified in their specialties. Academic labs, for example, are well positioned for basic research. Academic researchers are rewarded primarily based on their research performance, as measured by publication of cutting-edge science, and not by the commercial value of their outputs.

This focus on basic research makes academic researchers poorly positioned for applied research. Whereas academic researchers may be primarily interested in elucidating the mechanism of a biological interaction, a commercial researcher may be more interested in the potential to modulate an observed interaction, or in the potential for lead compounds to pass safety and efficacy tests in pre-clinical and clinical trials.

Stratification also exists within commercial research. Smaller companies generally lack the resources for large-scale manufacturing or clinical trials, leading them to focus on late-stage basic research and early-stage applied research, and transferring strong leads to larger companies for further development.

SPIN-OFF

Corporate spin-offs—creating a new entity to develop an opportunity—may be motivated by the ability of the spin-off to obtain financing unavailable to the parent, by the ability of the spin-off to absorb more risk than the parent company is willing to bear, or simply to develop promising leads shelved as a result of corporate restructuring.

By spinning off a promising technology or product to a newly-formed entity for development, an established company can maintain its focus on core technologies while the nascent company can obtain sources of funding that are unavailable to the larger company and assemble a specialized team to develop the spun-off technology.

> ### Box
> # Speedel: Spinning off to develop a shelved drug
>
> Speedel was formed in 1998, in the wake of the 1996 merger of Sandoz and Ciba–Geigy that formed Novartis, to focus on SPP100, a drug lead which was shelved by Novartis. Founder Alice Huxley, who had been working on the drug at Ciba-Geigy prior to the merger, convinced Novartis' management to spin-off Speedel.
>
> In licensing the drug lead to Speedel, Novartis retained a buy-back option, which they exercised in 2002. SPP100 received FDA approval and was launched as Tekturna in 2007. Besides being Speedel's first drug, Tekturna is also the first in a new class of drugs for high blood pressure, blocking kidney renin production (which can raise blood pressure).
>
> One of Novartis' primary motivations in abandoning SPP100 was concerns about the ability to manufacture the drug profitably. Speedel was able to overcome the high manufacturing costs by developing a new synthetic process. This arrangement shielded Novartis from the manufacturing process developmental risks and lilely granted Speedel scientists greater operational flexibility.
>
> Interestingly, Speedel later launched a dispute over claims that Novartis failed to pay them royalties on use of their manufacturing technology. In July 2008 Novartis announced plans to acquire Speedel for an estimated $880 million.

Spin-offs also permit the financial postings of parent companies to be shielded from the high negative revenues typical of start-up companies, enabling them to limit their exposure to the developmental risks inherent in biotechnology while still permitting them to benefit from the significant upside of new product development.

A European study found that corporate spin-offs are more likely to succeed than university-based start-ups. Commercial spin-offs demonstrated higher growth rates, a lower chance of failure, and often produced more innovations than independent or university based start-ups. Whereas university spin-offs showed a 45 percent failure rate, the failure rate for commercial spin-offs was only 15 percent.[1]

Factors attributed to the relative success of corporate spin-offs were greater business expertise and better access to capital and markets. Excessive bureaucracy and a less entrepreneurial approach were identified as factors impeding university start-ups. Another important distinction is that commercial spin-offs often result from restructuring, motivated by a desire to isolate research that does not match the core business or is too expensive. Because a company can always out-license or simply terminate projects that are not sufficiently appealing, the decision to spin off a technology may be an indirect indication of support from the parent company.

An additional factor impeding university spin-offs stems from the differing attitudes towards patenting in academia and biotechnology companies. The focus on frequent publication of research results in academic environments means that patenting is often employed primarily to protect intellectual property prior to publication. Conversely, most biotechnology companies place an emphasis on strategic patenting, resulting in patents designed to exclude competition, and which may be crafted to suit anticipated needs (see Box *Amgen v. Transkaryotic Therapies: Strategic patenting* in Chapter 7).

In an extreme example of the value of spinning off companies, Genentech's market capitalization increased from $11 billion to $16 billion in just six weeks as a result of being bought and then spun off in an initial public offering. In June 1999 Roche owned 65 percent of Genentech and exercised an option to purchase the remaining 35 percent, granting full ownership. Six weeks later Roche sold 20 percent of its Genentech holdings in an initial public offering and saw Genentech's market capitalization rise by $5 billion.

BUSINESS MODEL

Biotechnology companies can pursue research in one or more of the general categories of medical applications, agricultural applications, and industrial applications (see Table 10.1).

---

1 Moncada-Paternò-Castello, P., Tubke, A., Howells, J., Carbone, M. The impact of corporate spin-offs on competitiveness and employment in the European Union: A first study. *European Commission, Joint Research Centre, Institute for Prospective Technological Studies*, 2001.

Table 10.1 *Biotechnology application categories*

	Red: Medical biotechnology
Description	Drugs and other agents to treat, cure, or prevent disease, and products that assist in the diagnosis of diseases or measurement of critical factors in health and disease.
Characteristics	High up-front development costs, FDA (or other regulatory) approval required prior to sale. High post-approval profit margins.
	Green: Agricultural biotechnology
Description	Products and applications related to livestock and crop production, and agricultural production of biotechnology products.
Characteristics	Development costs are often similar to drugs, profits are often lower.
	White: Industrial biotechnology
Description	Modification or improvement of industrial processes or performing tasks previously served by industrial processes such as paper processing, bioremediation, or synthesis of organic compounds.
Characteristics	Reduced regulatory burden decreases development costs.

See chapter 6 for details on specific applications

Biotechnology firms are distinguished from pharmaceutical firms and firms casually using biotechnology techniques by their intense research focus and the emphasis on molecular biology techniques. While the prototypical biotechnology firm focuses on drug development, applications range in diversity from using bacteria to decompose oil spills to using genetically engineered bacteria to producing spider silk in goat milk (see Chapter 6).

Within the aforementioned application categories there are five basic activities in which biotechnology companies engage: basic research and target discovery; applied research and lead refinement; clinical and prototype research; manufacturing; and, sales and distribution (see Figure 10.1). Vertical integration was once a reality for pharmaceutical companies and a goal for emerging biotechnology companies, but a number of factors (see *Size* later in this Chapter) have made it more efficient for companies to specialize in just a few elements of the research-development-commercialization path. Most companies focus on just one or two of these activities, with a select few vertically integrated companies engaging in most or all of these activities. In the case of drug development, the costs associated with development and commercialization activities prohibit all but the largest biotechnology and pharmaceutical companies from integrating all the components. Even large vertically integrated companies outsource selected operations to smaller specialized firms.

Further dividing the above biotechnology industry segments, biotechnology companies can be segmented based on whether they focus on selling products or services. Nearly all biotechnology companies sell products and services to other companies, as opposed to selling directly to consumers. Products include physical items such as drugs, tools, reagents and other manufactured compounds. Service firms perform defined services rather than selling defined products. Examples of services include research support activities such as contract research, manufacturing, lead-optimization, and diagnostic services such as paternity and forensic testing.

Some companies combine product development and service offerings in a hybrid model, selling proprietary services while using

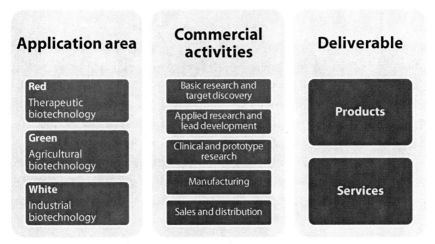

Figure 10.1 *Biotechnology company activities*

them internally for product development as well. This strategy can potentially demonstrate the value of a company's scientific foundations and defray the high costs of product development, but it is also subject to abuse. Companies with faltering core elements may adopt a hybrid strategy to leverage their intellectual property and potentially distract investors; platform and service firms can add a drug development unit to distract investors from technologies facing obsolescence, and product-development companies with failed leads can re-deploy their resources by selling services and technologies based on their proprietary techniques and knowledge. A crucial question to ask of hybrid firms is whether their business model makes sense as part of a long-term strategy, or is potentially being implemented primarily to buy time or attract additional investors.

PRODUCTS AND SERVICES

While the first biotechnology firms focused on research and early-stage development, in just a few decades business plans have changed from capital-intensive pure research and development to more stable strategies with internal revenue streams.

Drug development is currently the favored application of biotechnology. The great potential of drug development stems from three factors. First, therapeutic products have the potential to treat large sustainable markets comprised of motivated buyers. Second, the financial and time investments required for drug development, combined with a strong patent portfolio, can create a significant barrier for competitors. Third, the FDA review process represents an objective series of milestones that companies can use to represent the progress of projects in development. This helps investors rationalize company valuations and creates market opportunities to sell drug leads that have passed clinical trial phases (see Figure 12.1 and Box *Clinical trials provide valuation milestones* in Chapter 12).

The sum of these factors facilitates the development of products with high profit margins and minimal pricing pressure; at least until patents expire. A significant downside for drug development is the need for extensive upfront investments to perform clinical trials, and the possibility of not gaining regulatory approval after these signifi-

Table 10.2 *Service payments*

R&D fees	Payment for research and development activities. May not require specific milestones.
Milestone	Payments attached to specific deliverables such as clinical trial progress or producing a defined product.
Royalty	A percentage of product revenues.
Service charges	Simple payment for services rendered.

cant investments. The legacy of strong returns realized by pharmaceutical companies stands testament to the profit potential of drug development.

Comparisons of the relative value of drugs and non-therapeutic biotechnology products must consider the development costs and financial rewards. Applications such as functional foods, for example, are likely to be subject to the same regulatory burden as drugs, but their profits cannot compare with drugs. Therefore, while these applications may have significant sales potential, their development is nonetheless discouraged.

Some of the research services that biotechnology firms provide are in developing and refining lead compounds, manufacturing developed products, and testing products. Research service firms can derive revenues from research fees, milestone payments, royalties, and service charges (see Table 10.2). Research fees compensate these firms for the services they provide. These may be paid as lump sums or dispersed as milestone payments, granted upon completion of defined phases of development. Royalty payments award a percentage of sales stemming from services provided. Service charges such as manufacturing and testing fees are distinguished from research fees in that the procedures and end products are better defined. Direct service firms, providing discrete services such as paternity or forensic testing, sell their services directly to end-users and are compensated by simple service charges.

TEAM

Small biotechnology companies have three basic team components: scientists conduct research, management provides direction and administrative support, and a scientific advisory board provides an objective assessment of the company's progress and goals. Larger companies may deploy additional components such as sales forces and legal teams, but as research and development are at the core of commercial biotechnology, the three aforementioned components are necessary elements in any biotechnology company.

SCIENTISTS

Scientists are the primary producers in biotechnology companies. It is the scientists who ultimately develop any products and technologies that lead to commercial success. It is essential to employ scientists with the specialized skills and experience necessary to develop and carry out research programs that lead to commercial success. Despite the availability of computers to collect and analyze data and sophisticated methods to simplify research, skilled hands and minds are still required to carry out and design appropriate experiments.

There are three basic roles performed by scientists in biotechnology companies: research support, research, and research management. The roles and responsibilities of scientists vary with their education and experience. While some companies may contract out the entirety or portions of their research, the need for quality scientists is not eliminated—it is merely shifted to contractors.

Laboratory technicians perform research support. These individuals clean glassware, maintain stocks of reagents, and perform routine research tasks and other functions required to keep labs functioning. Laboratory technicians may also perform supervised research experiments, and may or may not have college degrees.

A variety of individuals are directly involved in research. Depending on their education and experience, these individuals may be called research assistants, research associates, or research scientists. Research associates typically have a college or post-graduate degree and laboratory research experience. They perform scientific experiments under the supervision of senior scientists. Associate scientists have Ph.D. degrees with post-doctoral training. These independent investigators may or may not work on team projects, but they are still supervised by more senior scientists.

The role of directly managing research is left to senior scientists, research directors, and group leaders. With a Ph.D. and proven research expertise, these individuals are well situated to plan and manage multiple research projects and run research labs.

As described later in this chapter, the primary factor in biotechnology site selection is access to a skilled workforce. Specific research areas and manufacturing techniques require scientists with specialized skills. This granularity in personnel requirements challenges recruitment and makes employee retention especially important.

The reliance on skilled workers means that personnel shortages in biotechnology can be especially detrimental. Without the ability to hire workers with required skill sets, companies may be forced to outsource operations, train existing employees or new hires at significant expense, set an unsuitable worker to a task, or simply abandon a research task.

MANAGEMENT

Customers do not buy innovative research. They buy useful products and services. The ultimate goal of biotechnology companies is not to produce scientific advances; it is to profit from innovation. Management is responsible for identifying commercial opportunities and positioning a company to exploit them.

The actions of the management team in guiding the development and marketing of a product are essential to commercial success. It is important that management remain focused on satisfying customer needs. The research conducted by biotechnology companies focuses on developing innovations, but the ultimate goal of these companies is to profit from these innovations. While a good management team can sell a mediocre product, even the best product will not sell itself.

An effective management team must identify commercial opportunities and guide the company to realizing them. Because understanding the competitive landscape and commercial possibilities are essential for successful strategic guidance, managers with extensive industry experience are best suited to this task. In addition to the chief executive officer, scientific advisory boards and boards of directors are other sources of strategic guidance.

It is vitally important that the management team include someone with proven scientific expertise, often called chief scientific officer, chief technical officer, or chief medical officer. These are often separate roles in larger companies, but may all be filled by the same individual in small companies. The basic responsibilities of these scientific roles are to ensure that a company's scientific endeavors are aligned with corporate goals and to manage technical or clinical trials to ensure that products meet market and regulatory demands.

The role of an operational manager is to monitor the activities of a company to ensure that milestones are met and the overall strategic plan is accomplished. In small companies, operations management may fall under the domain of one of the above managers, but as companies grow a dedicated staff member may be added to optimize operations and relieve the other managers from the burden of day-to-day management.

The long horizon for sustainable revenues and the significant cost of innovative scientific research requires external sources of funding. An effective management team must contain a member, designated the chief financial officer, who understands the various forms of funding available for biotechnology development and is able to help raise funds. Additionally, the chief financial officer must manage the rate at which a company spends its cash reserves, the burn rate, to ensure that operations can be sustained until the company is able to accomplish its goals.

BOARD OF DIRECTORS

The board of directors of a company is a group of individuals legally charged with the responsibility to protect the interests of the company and its shareholders. While employees ultimately answer to the chief executive officer, the CEO must answer directly to the board of directors. The board of directors is also responsible for determining a company's mission and purpose, selecting a chief executive officer, and monitoring executive and corporate performance.

As representatives of the shareholders of a company, the board of directors represents a means to override the CEO and managers in the event that their actions and decisions conflict with the interests of shareholders. Such a conflict might occur when a third party seeks to acquire a company and potentially liquidate its assets. While management may resist the action, seeking to retain their employment, they may be superceded by the board of directors in the greater interest of the shareholders. The board of directors can also

---

Box

Shareholder activism

Public companies are owned by their shareholders, and it is the job of management to deliver value to their shareholders. The board of directors serve as representatives of the shareholders, and have the power to overrule company managers. Companies that fail to satisfy shareholder demands may see board members, directors, or CEOs replaced. Shareholders can also oppose unfavorable acquisitions, or press a poorly performing company to liquidate and return proceeds to shareholders.

A high profile example of shareholder activism was Carl Icahn's role in the sale of MedImmune. Icahn, perceiving MedImmune as a company that would deliver more value in an acquisition than by continuing to operate under its existing management, purchased 1.16 percent of the company in early 2007. Echoing the sentiments of existing shareholders, he threatened a proxy war—a process by which individual shareholders can vote for corporate changes—to grant him a board seat if MedImmune did not agree to seek a buyer.

Since having a board seat would have given Icahn the power to directly push for a sale, his threat was sufficient to compel management to seek acquisition. Several weeks later, MedImmune reached an agreement to be acquired by AstraZeneca for $15 billion, a 50 percent premium over MedImmune's price prior to seeking a buyer, granting Icahn an estimated $300 million profit.[1]

---

1 Hamilton, D. Icahn scores lucrative victory in MedImmune sale. *Reuters*, April 23, 2007.

be held responsible for malfeasance, and must therefore monitor management to ensure that improprieties are not occurring.

Beyond their vital role in protecting the shareholders of a company, the board of directors also makes important contributions in setting the direction of a company. Whereas the management team *manages* a company's operations, the board of directors *directs* the company's strategy. Accordingly, it is essential that a board of directors include members who are intimately familiar with the scientific, regulatory, and commercial elements relevant to a company's operations.

Board members can also complement management gaps in growing firms. An effective board includes members from a broad range of disciplines relevant to a company's activities. A drug development firm's board, for example, should include expertise in clinical trials, reimbursement, and marketing. These experts may be tapped to fill short-term needs such as helping design clinical trials or negotiating reimbursement, avoiding the need to hire a specialized manager.

Scientific Advisory Board

The scientific advisory board (SAB), intentionally removed from management, is tasked with guiding the scientific activities of biotechnology companies. It is critically important, especially for young companies, that the board be objective. This diverse group of individuals with experience in the industry should understand the overall view of the system being studied and must have the capacity to direct the company on how to approach scientific problems and determine which routes are likely to provide a solution to the problem at hand.

Scientists working for biotechnology companies are spared the need to compete for research grants. This administrative benefit is balanced by the absence of the pressures and challenges of periodic justification of research efforts. An important role for the SAB is to provide an alternative to the peer-review panels that review grants for researchers in academic institutions. As a source of objective judgment of the direction and quality of a company's research as well as evaluation of the effectiveness of scientists, the SAB is a vital part of a biotechnology company.

SAB members are usually academic scientists who are highly regarded in their field and may be well known in business and investment circles as well. It is essential that they have extensive experience in scientific or clinical areas pertinent to the startup. Individuals may be motivated to join SABs for a variety of reasons: they may have a desire to contribute their expertise; they may be interested in a company and wish to be involved in its development; they may be interested in licensing opportunities; and, they may want an equity stake in a promising company.

As a company matures, the board should also change to adjust to the changing needs of the company. A young company can benefit from a small board that is focused and can make decisions quickly, while a more mature company can benefit from a larger and more diverse board where more voices can participate in making strategic recommendations.

In addition to guidance, companies seek out SAB members for two other purposes. SAB members are desired who will add credibility to a startup, or who can attract talent and funding. The inclusion of a Nobel Prize recipient or other esteemed scientist serves as an indication that an experienced and knowledgeable individual sees promise in a company. SAB members may also be instrumental in recruiting workers and attracting funding for a company by tapping their own laboratories and personal networks.

COMPANY CHARACTERISTICS

The objectives of biotechnology companies and the challenges they face are reflected in their design and business structure. Biotechnology business development requires three elements: research proficiency to enable development, funding to support development, and a competitive advantage to enable profitable commercialization. It is the role of management to bring together these necessary elements and guide commercial activities to a productive end. A company's location, size, and maturity are other important characteristics that influence business operations.

LOCATION

In selecting a business location, biotechnology companies must consider factors such as access to capital and skilled workers, costs of doing business, regional laws, and regulations. The accumulation of biotechnology in an area enhances access to skilled workers and capital and leads to the development of biotechnology clusters.

The influencing factors in biotechnology company site selection change as a company matures. Start-up companies focus on research and development and are therefore more motivated by factors enabling research, such as access to skilled workers and funding, than by tax credits or the cost of research space. Mature companies with sustainable operations are able to attract capital and employees from distant locations and accordingly place more emphasis on factors affecting their bottom line, such as operational costs.

Zoning is also a significant issue. Biotechnology companies may work with dangerous chemicals, radioactive materials, or perform animal testing. The inability to receive permits for these critical elements can be a non-starter in location decisions.

One of the most important needs of young biotechnology companies is access to skilled workers, who tend to aggregate in areas with high concentrations of biotechnology companies or academic research laboratories. Because biotechnology research is risky by nature, employees must consider their options in the event that their job is terminated. Employees of biotechnology companies often have specialized expertise, which makes the difficulty of finding a new job in the event of downsizing or restructuring an important consideration. Having several potential employers in a region, such as biotechnology companies, service providers, and universities, can ease transitions for employees. Biotechnology companies without nearby neighbors may therefore find it difficult to recruit top-flight researchers.

The need to offer workers career security causes biotechnology start-ups to congregate in close proximity, often near sources of skilled workers. Because the best places to find top workers are areas with high quality universities, medical centers, and research centers, it is not surprising that many biotechnology companies have formed in the vicinity of high concentrations of academic institutions and medical and research centers.

Factors discouraging location in clusters are increased cost of living and the competition for workers and funding. The valuation of biotechnology companies is also inflated in biotechnology clusters, partially due to increased competition among investors to

---

Box

Hybritech's crucial role in San Diego's biotechnology industry

Hybritech was formed in September 1978 by Dr. Ivor Royston, an untenured UCSD professor, and his research assistant Howard Birndorf, with the objective of commercializing monoclonal antibodies. Hybritech was one of the first biotechnology firms, and is credited with helping lay the foundations for San Diego's biotechnology cluster.

Why did Hybritech form in San Diego? The simplest explanation is that Royston was a UCSD professor and started his company close to home. With $300,000 in seed funding, Hybritech set its focus on diagnostic tests and went public in 1981. As the company grew it started to attract attention from pharmaceutical companies. Following the 1985 development of a test for prostate cancer, Eli Lilly purchased Hybritech for nearly $500 million. A collision of cultures quickly ensued between Hybritech, whose managers and scientists were accustomed to a very casual corporate structure, and Eli Lilly's legacy pharmaceutical company corporate structure.

Within a year of the acquisition, most of Hybritech's key talent had left the firm. Many of these individuals, rich in experience and cash, went on to start new biotechnology ventures or became venture capitalists and funded new biotechnology ventures. With the support of community organizations and an entrepreneurial environment San Diego has grown to become the third largest biotechnology cluster in the United States, after the San Francisco Bay area and Boston. More than 100 San Diego companies can trace their history to Hybritech.[1]

---

1 Local life science industry breeds serial entrepreneurs. *San Diego Daily Transcript*, October 19, 2007.

purchase equity in promising projects. If a source of well-educated workers is present outside of a biotechnology cluster, local companies may be able to secure employees at a discounted price relative to cluster-based competitors due to reduced competition for talent and the desire among some individuals to stay close to their homes and extended families. Because of reduced competition for funding, it may also be easier to secure funding. This is countered by the reluctance among many investors to travel far from their home bases. One method to support development in the relative absence of early-stage investors is to form regional biotechnology development initia-

---

Box

Research Triangle Park

While biotechnology clusters may form organically in the vicinity of strong research bases, government actions can also facilitate development. A prime example of a planned biotechnology cluster is Research Triangle Park, in North Carolina.

In the 1950s North Carolina suffered from brain drain and a declining economy. Markets for the state's primary industries—textiles, furniture, and tobacco—were declining and college graduates typically left the state to seek work elsewhere. Seeking to capitalize on the three local universities, Duke, North Carolina State, and the University of North Carolina, local politicians and businesspeople sought to create local opportunities to stem the brain drain. Research Triangle Park was founded in 1959 on 4,400 acres of worn-out farmland.

Following a slow early start, years of lobbying government and private institutions started to pay off in the late 1960s and 1970s as industry leaders and government agencies decided to locate laboratories and offices in Research Triangle Park. Common draws to locate in Research Triangle Park were the quality of life and modest costs. Today Research Triangle Park encompasses nearly 7,000 acres, is host to more than 150 organizations, and its 39,000 employees draw more than $2.7 billion in salaries, making it the largest planned research center in the world. Roughly one third of the resident firms and organizations are biotechnology and pharmaceutical companies.

tives. These groups can offer incubator facilities and other resources to support biotechnology startups until they can attract sufficient capital to progress independently.

Relative to younger companies, developed companies with established reputations are more able to recruit distant talent and capital, granting them greater flexibility in their choice of location. The needs of mature companies emphasize the ability to sustain their operations. With an established business, regional influences on the cost of doing business can be objectively assessed and may have a significant impact on a company's bottom line. Mature companies must consider the availability of medical professionals for clinical trials, land availability, and local regulations such as animal testing bylaws. State governments and local municipalities may also offer economic, financial, and training incentives.

FACILITIES

The facilities where biotechnology research is conducted have special design constraints. There are several important considerations from a real estate developer's point of view. The cost of developing biotechnology facilities can easily exceed that of general office

Table 10.3 *Biotechnology facility operating costs*

	San Francisco, CA metro area	Chicago, IL metro area	Philadelphia, PA metro area	San Antonio, TX metro area
Labor	$14,811,216	$13,802,640	$13,462,640	$12,109,984
Electricity and natural gas	471,540	345,660	468,720	224,160
Amortization	5,422,114	3,061,452	2,545,645	1,726,852
Property and sales tax	1,436,921	1,414,134	1,306,846	1,169,062
Lease	3,264,000	2,245,000	2,448,000	2,030,000
Heating and air conditioning	123,470	280,000	348,272	125,062
Corporate travel	149,848	161,284	149,428	109,536
Total	$25,679,109	$21,310,170	$20,729,551	$17,494,656

Estimates are for a 200 worker, 100,000 sqft facility
Source: Bizcosts.com

> **Box**
> # Chiron's flu vaccine facility loses license
>
> In 2004 the Medicines and Healthcare Products Regulatory Agency, the British counterpart to the FDA, and the FDA revoked licenses for Chiron's Liverpool plant to sell flu vaccines. The plant had identified contamination problems in several vaccine lots but failed to adequately address them. The fallout of the resulting shortage of flu vaccine supplies had far-reaching consequences.
>
> The United States had been expecting to have 100 million doses of flu vaccine in preparation for the 2004/2005 flu season, with 48 million doses coming from Chiron's Liverpool plant. The withdrawal of Chiron's vaccine led to rationing and price increases for consumers, and a significant financial penalty for Chiron. Prior to the debacle Chiron enjoyed a roughly 50 percent share of the U.S. flu vaccine market. Chiron's stock fell by 30 percent in the wake of the plant closure, and earnings dropped by half. Other companies quickly moved in to fill the void left by Chiron, challenging Chiron's potential for recovery.

space by an order of magnitude, so developers need assurance that they will have sufficient occupancy to recoup their investments. Additionally, conversion from biotechnology space to other uses is not a simple task. The prior use of animal research subjects, dangerous chemicals, and radioactive materials may discourage potential future tenants. Many universities and communities have developed incubator facilities which offer laboratory space to start-ups at reduced costs in order to overcome reluctance among developers to invest in biotechnology research space and to assist start-ups encumbered by the high cost of leases.

The physical requirements for R&D facilities depend on the activities performed and the level of regulatory scrutiny required. Most laboratories require high electrical power densities, vibration-resistant floors with high load carrying capacities, purified water supplies, gas lines, robust climate control, and controlled airflow systems. Electrical systems must also provide extra protection against surges that can destroy expensive equipment. Biotechnology research is dependent on the use of freezers to preserve biological samples; some samples are kept in special very low temperature freezers for decades. Losing the contents of a freezer in a power outage can destroy the products of years of research, necessitating redundant power backups.

Dangerous and volatile chemicals are ubiquitous in biotechnology research, necessitating special fume hoods, extensive air recirculation, enhanced fire suppression, and other safety systems. Cleanrooms and animal facilities also require validated air and water purification systems to prevent contamination. FDA Good Manufacturing Practices (cGMPs) regulate manufacturing facilities and have special validation requirements in a number of additional areas, extending as far as water piping conventions. Failure to meet and maintain these stringent requirements can result in loss of facility approval, potentially costing millions of dollars in lost sales and denying patients essential medications.

SIZE

Nascent biotechnology firms tend to be relatively small and require external funding to sustain research until they are able to independently generate sufficient revenues to maintain operations. These nascent firms often lack manufacturing, sales, and distribution abilities as well, precluding the capacity to market any products they may develop. Lacking sufficient bulk to enable wide-scale commercialization, these companies often license products or partner with senior firms to sell products. Mature firms tend to be relatively larger and operate independently using revenues from sales and licensing. These firms often leverage their late-stage development and commercialization abilities by obtaining innovative products from smaller firms through development partnerships, by purchasing products, or by licensing product rights. Licensing, alliances, and other inter-company activities are described in Chapter 14.

While it is commonly assumed that larger research and development programs produce disproportionately larger returns than small ones, this has not held true for pharmaceutical companies.[2] Despite

2 DiMasi, J.A. New drug innovation and pharmaceutical industry structure: trends in the output of pharmaceutical firms. *Drug Information Journal*, 2000. 34(4):1169-94.

consolidation in the pharmaceutical industry, new drug approvals for the four largest firms decreased by 30 percent in the 1960s and by 18 percent in the 1990s. A study by McKinsey & Company found little long-term correlation between size and returns to shareholders based on 1994 drug revenues.[3] This implies that small firms are not necessarily at a disadvantage due to their size; it is possible for a small company's R&D program to develop high-value products.

Gross productivity trends aside, there are certain characteristic benefits and challenges associated with size. Small biotechnology companies tend to only have sufficient resources to pursue a relatively small number of projects. This means that they cannot spend too much time or too much money investigating potentially lucrative side projects. A significant challenge for these small firms is that they may reach a hurdle in the course of development that they are unable to surmount due to a lack of expertise, funding, or time. Additionally, because of the uncertainty of the scientific potential of their research projects, the quality of a small company's team tends to be more important to investors and partners than the quality of the science—individual research projects may succeed or fail, but the quality of the team is an important, and measurable, element.

Compared with small firms, large biotechnology companies can devote more resources to research areas that look promising or require extra support. The ability to perform critical proof-of-principle research quickly is especially valuable for public companies, because if failures happen early enough they do not have to be reported. Larger companies are also better able to divert resources to interesting sidelines, enabling them to capitalize on opportunities that would be lost by smaller firms. While a failed project can spell disaster for a small firm, larger companies can use income from established products to buffer the impact of such disappointments.

Just as investors diversify their investments to reduce risk, a company developing multiple products is less dependent on the success of individual products. For example, with two projects it is possible to discontinue the less-promising project to focus on the more-promising one. A company with only one project is not likely to have the resources to start a new project should its sole project

3 Agarwal, S. *et al.* Unlocking the value in big pharma. *The McKinsey Quarterly*, 2001.

lose appeal. This lack of options may lead investors to readily accept, deserved or not, suggestions that a smaller company is continuing development of a project that is doomed to fail, out of lack of any alternatives. Larger companies can at least prioritize their goals, giving investors a relative appreciation of potential success. Therefore, small size combines the problems of increased risk due to lack of diversification with exclusion from consideration by some investors.

The ability of small companies to focus on specific research areas or applications can provide them with a significant advantage over larger and more diversified competitors in niche markets. Smaller size and a decreased bureaucratic load also mean that smaller companies can be more agile than large ones. This agility is often combined with a lack of experience and expertise. Therefore, while smaller companies can make decisions faster than large ones, they may not make the right ones.

Vertical integration of all components from discovery to commercialization was once a reality for pharmaceutical companies and a goal for biotechnology companies; this is no longer the case for either. The rate of change in technology and markets, and the number of potential allies and competitors favor flexibility. An established company that can form alliances with several research partners can gain from the specialized expertise of individual partners. Furthermore, outsourcing elements of development can effectively minimize risk by enabling a company to take a stake in several promising research projects. The flexibility gained by alliances and outsourcing provides expanded access to innovative technologies and permits the termination of discouraging external projects with greater ease than internal projects.

Beyond licensing and alliances, large companies may also employ directed strategies to drive innovation. Examples of strategies used by large companies to expand their innovation capacity are described in the section *Outsourcing Innovation* in Chapter 12.

MATURITY

In biotechnology, stability comes with size and maturity. Biotechnology companies can be roughly divided into three groups

Table 10.4 *Assessing biotechnology company maturity*

	Profits	Products
Mature	Measurable and significant	Products on market and in development
Promising	Small profits or strong prospects for near-term profitability	Products on market, emphasis on products in late-stage development
Emergent	No clear path to profitability	Developmental uncertainty

according to these qualities. While the distinction between these groups is somewhat subjective, it provides a useful framework to distinguish between the various stages in the maturation of a biotechnology company.

Mature companies are established large-cap biotechnology companies and have relatively stable revenue streams; promising companies have strong fundamentals and excellent prospects for near-term profitability; emergent companies are nascent biotechnology companies without a clear path to profitability, making it difficult to reliably assess their long-term potential.

Mature companies are profitable. Compared with promising and emergent companies, the revenues that these companies earn permit them to invest relatively more resources in research and development programs. While they still face the inherent risks of biotechnology product and service development, mature companies will fare better than smaller ones in the event of a research disappointment or in unfavorable market conditions. These relatively stable companies lend themselves to traditional financial analysis far more readily than promising or emergent biotechnology companies.

Promising companies either have products and services on the market with growing sales, or have potential big sellers in late-stage development or clinical trials. A well-positioned promising company may be unprofitable, but it will have sufficient cash reserves to fund research and operations until profits emerge. Some of the companies in this group will go on to become industry leaders while others may fail, remain small, or be acquired by competitors. Promising companies involve significant risk.

Emergent companies are still seeking to commercialize their first product or service or are struggling to gain acceptance for their technology platforms. These companies may announce that they have products in development which address lucrative market opportunities, but on closer examination it often becomes clear that these products are years away from completing development. Aside from the risk of developmental failure, a second risk is that competitors may emerge or market trends may shift and these emerging companies will be in the uncomfortable position of having devoted their resources to developing a product for which the market potential has diminished. Eventual success for these companies is far less certain than for promising companies.

Chapter 11

Finance

Where else can you combine making money—which is the primary purpose of venture capital—with the feeling that you're doing some good in the world?
Venture capitalist Frederick Adler, regarding biotechnology

Biotechnology companies employ a variety of methods to fund the expensive research and development programs that are necessary to produce marketable products and services. Sophisticated machines and well-trained workforces cost a lot of money. The requirement for precision and consistency demands refined tools and reagents of great purity which are expensive to purchase, maintain, and operate. The cost of attracting and retaining the highly skilled workers necessary for cutting-edge research is also significant. The research and development required to produce biotechnology products and services takes a long time and is fraught with unexpected setbacks, requiring substantial quantities of funding long before revenues can be anticipated.

These high capital requirements, combined with uncertainties about market size and projected revenues, and the relative lack of tangible assets to use as collateral, prevent the extensive use of conventional debt financing, particularly for biotechnology companies that do not have a proven product. *In lieu* of debt, biotechnology companies often exchange equity with professional financiers or sacrifice some of their autonomy to sponsoring partners in order to finance research and development.

Funding for biotechnology companies comes from various sources. Professional financiers exchange money for equity in promising companies with the hope of realizing a return on their investment. Established companies can spin-off new entities to exploit promising research projects, contract elements of research to firms with specialized expertise, or form partnerships with innovative companies and provide cash in exchange for research leads or the right to market developed products. Various federal, state, and regional programs also provide funding opportunities to biotechnology companies. Industry estimates of the sources of capital for the first decade of a biotechnology company's existence are that 10 percent comes from venture capital and other private equity sources, 40 percent from public markets, and 50 percent from senior partners.[1]

Just as a public company's stock price represents the market's assessment of that company's value, the caliber of a start-up's financial and commercial partners are an excellent measure of that firm's long-term prospects. Funding from a top-flight venture capital group or angel investor with proven industry expertise stands as testament that experienced industry insiders have high aspirations for a company. Likewise, partnerships with industry leaders serve as an endorsement of the quality of a company's research and development or products. This esteem-by-association is called institutional support.

DEVELOPMENT STAGES AND FUNDING

A developing biotechnology company will go through several discrete stages during its growth. Regardless of a company's origins— whether it is spun-off from a larger firm or formed to capitalize on technology transfer from academic labs (see *Technology Transfer* in Chapter 10)—the same stages exist. In some cases, early funding requirements may be met through the sponsoring institution prior to the actual founding of the nascent company.

The general framework described below considers a product development company, although the discussion is relevant to service companies as well. While this presentation makes the process appear linear, it seldom is. Biotechnology companies need to continually investigate lucrative tangents emerging from research, to si-

---

1 Hess, J., Evangelista, E. Pharma-biotech alliances. *Contract Pharma*, September 2003.

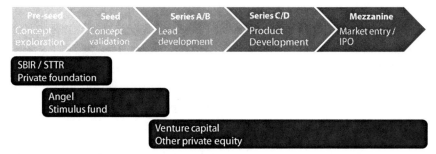

Figure 11.1 *Funding stages and sources*

multaneously pursue multiple opportunities, and to adapt business and research plans to accommodate changing internal and external influences.

Funding is commonly tied to discrete milestones, which should provide objective measures of progress toward commercialization and profitability. The completion of milestones provides evidence that the goals are attainable, reducing risk and thereby increasing the valuation.

In the seed stage, a biotechnology company simply has an idea for a product. There is an immediate need for sufficient funds to develop a proof-of-principle so that the company can attract further funds and develop a prototype. The funds that launch a company, fund its proof-of-principle research, and generally support an early assessment of business feasibility, are called seed funds. This initial stage is generally the most dilutive, with financiers commonly seeking ownership of 50 percent or more of the company (see Table 11.1).

While venture capitalists once commonly provided seed investments, seed stage funding is now more commonly provided by angel investors, friends and family, and bank loans (which may be personally guaranteed by the founders). As a company develops, each financing stage is generally less dilutive than the previous stages. Within a short time after seed funding, generally six months, a company should be able to demonstrate proof-of-principle. This often coincides with patent filing, which is the first step to securing market protection for the product in development.

The expansion stage involves production of prototypes and requires more money than the proof-of-principle stage, granting a relatively smaller share of equity for the amount of money invested (see Table 11.1). This stage may see several rounds of funding, often named by how many external fundings have occurred. The seed round is named first-stage. Series A, series B, and subsequent rounds are named accordingly. These post-seed funding rounds may each raise millions of dollars and should support company operations until commercialization occurs.

The final private funding stage, mezzanine funding, is used to prepare products for market and enable a company to independently survive on revenues and loans, or to launch an initial public offering (IPO). As an alternative to public offerings, companies may be acquired by other private or public firms (see Box *Alternative route to going public* later in this chapter).

HOW MUCH MONEY SHOULD YOU RAISE?

Because of the characteristic high capital requirements and inherent developmental risk in biotechnology, funding is seldom provided as a single lump sum. Financing rounds and milestone payments provide companies with incremental funding as development goals are met.

Common guidance dictates that a company should raise as much money as it can; a growing company should always be seeking funding, and should accept as large an investment as investors are willing to make. The basis for this strategy is that one never knows when markets will turn sour or when investor interest (or their own liquidity) will wane, and should therefore capitalize on every opportunity to raise money.

Others maintain that it is essential to develop a balance between excessive funding that can impede growth and insufficient funding that can threaten a company's existence. Not raising enough money to reach a milestone makes it difficult to entice new investors, often requiring a company to return to previous investors for more money. This sign of weakness may lead investors to seek relatively greater amounts of equity. Conversely, raising too much money in early rounds may provide a company with an undeservedly high valuation or promote an inappropriately high burn rate (the pace at

Table 11.1 *Progressive equity dilution by funding stage*

Funding stage	Investment	Percent ownership
Seed	$300-600k	40-60
First-round	$1-5 million	40-60
Early mezzanine	$5-15 million	20-30
Late mezzanine	$20-50 million	25-35

Source: Birndorf, H.C. Rational financing. *Nature Biotechnology*, 1999. 17:BE33-BE34

which an unprofitable company spends its cash). Both these attributes can dissuade potential investors looking for companies poised for dramatic growth. Furthermore, because earlier financing rounds are more dilutive than those that follow—they exchange relatively greater proportions of equity for cash—founders may find their equity shares significantly diminished in later rounds.

EQUITY DILUTION

An important concept to understand in raising capital is equity dilution. Investors put capital into companies in exchange for equity—a share of the company—and founders, employees, and early-stage investors alike must be aware of the impact of dilution of their ownership as a company grows.

When founders have an idea for a company, they own 100 percent of the equity in the future company. The current value of this company, however, is effectively nothing. Capital needs to be raised to incorporate the company, to fund R&D, and to develop the commercial ideas into marketable products. This capital is exchanged for equity in the company, so the founders may be left with only a small portion of equity by the time a liquidity opportunity emerges, enabling them to sell some or all of their shares. Early-stage investors, who may have obtained 40 to 60 percent of the company's equity for their initial investments, will likewise find their equity diluted by downstream investors. Table 11.1 shows a general model for progressive equity dilution by funding stage.

It is worth noting that the habit of downstream investors to 'bid up' the value of a company increases the absolute value of shares purchased by prior investors, despite the dilution of their ownership. Early investors may also use anti-dilution provisions in term sheets, enabling them to maintain a constant ownership share, or at least to diminish their dilution, as new shares are issued.

PRIVATE EQUITY

The types of individuals and groups engaged in financing a biotechnology company change as the company grows. These investors vary in their tolerance for risk, their expectations for returns, and their financial resources. Figure 11.1 shows the general alignment of funding stages with funding sources. Private equity investors, described in this section, make investments in promising private firms in exchange for a share of equity, or ownership.

Investors can be divided into two groups based on their ability to contribute to a company. In addition to capital, financiers and corporate partners may also contribute institutional support, guidance, and networking contacts. "Dumb" money comes from investors who lack the ability or inclination to provide advice or networking. "Smart" money comes from investors who can contribute advice and business contacts. Additionally, an important distinction between professional investors and corporate partners is that corporate partners are often motivated by product or service delivery rather than a simple financial return on investment. Accordingly, the level of support they provide may be far more valuable than monetary contributions. Conversely, corporate partners may demand the option to purchase a company or license products at pre-determined prices or formulas, limiting profit potentials to shareholders.

FRIENDS AND FAMILY

While seed stage funding may be obtained by offering equity to friends and family or seeking loans, later stages require professional financiers who are able to raise larger funds, assess risk, and provide guidance and other services. These professional investors include angel investors, venture capitalists, merchant banks, investment funds, incubators, and corporate partners who finance and share business expertise and contacts with entrepreneurs.

With previous experience in developing or funding biotechnology companies, financiers can help growing companies avoid pitfalls. Moreover, unlike other sources of guidance such as consultants, professional financiers have a long-term vested interest in a company's success. This long-term interest is exemplified by their compensation. Rather than charging for their services, the guidance and assistance of professional financiers are sometimes viewed as justification for the opportunity to invest in a company. The prestige and networking contacts associated with partnership with esteemed financiers can also facilitate additional funding, help secure top researchers and managers, and attract representation by leading investment banking and law firms.

ANGEL INVESTORS

Angel investors are wealthy individuals who invest in private companies. According to Securities and Exchange Commission guidelines, angel investors must be accredited investors; they must have an individual net worth exceeding $1 million or earn more than $200,000 annually (or $300,000 in combination with their spouse) to be accredited. Angel investors tend to have succeeded in their field and are looking for opportunities to help young entrepreneurs develop new companies. A significant difference between angel investors and venture capitalists is the scale and intensity of their activities. Unlike venture capitalists who manage venture partnership-raised funds, angel investors invest their own money, and traditionally make investments with less-onerous terms.

Because helping companies grow is usually not their full time occupation, angel investors often invest in fewer companies than venture capitalists and may spend less time actively involved in company development.

One characteristic by which angel investors differ from other financiers is that they often focus on their core interests independent of market sentiments. This may give angel investors an edge over larger investors because they can develop positions in firms developing applications with unrecognized potential. Furthermore, the resources needed to perform due diligence for a nascent biotechnology company are comparable to those needed to evaluate more mature firms, motivating many larger investors to ignore the smallest biotechnology firms. This void frees angel investors to select small companies for investment, enabling them support the earliest business development activities, such as proof-of-principle, preparing companies for investors with greater resources who will support further growth.

VENTURE CAPITALISTS

Venture capitalists are professionals who manage venture capital funds. Their goal is to invest money in promising companies in return for equity. Experienced in developing companies, venture capitalists offer their business expertise in exchange for the opportunity to invest in a company. In a competitive environment, venture capitalists may compete for the opportunity to invest in individual companies.

Venture capitalists often limit their investment criteria based on commercial focus, geographical location, and maturity. While some venture capitalists in biotechnology hubs do not invest in companies located outside of their immediate area, others are willing to travel outside of biotechnology hubs, seeking exceptional companies with relatively lower valuations than local firms.

The investors in venture capital funds include wealthy individuals and corporate entities such as pension funds. In order to produce the return on investment that their investors seek, venture capitalists make a number of risky investments in the hopes that at least

---

Box

Venture capital: The poker analogy

A lot of people think that venture capital is the most risky form of investing there is. Certainly the failure rates are higher than any other form of investing I know of. But I believe that venture capital is much less risky than people realize. And that's because we have the opportunity to scale into our investments, to "dollar cost average."

And that leads me to the poker analogy.

Poker is an incredible game. It is about risk management and knowing when to go for it and when not to. So is the VC business.

Early stage venture capital is a lot like poker. The first round is the ante. I think keeping the ante as low as possible is a good thing. I like to think of it as an option to play in the next round and to see the cards. Clearly, we don't ante up to just any deal, but it is very useful to think of the first round as the ante.

For the first year or 18 months, however long the first round lasts, you get to "see your cards." You learn a lot about your management team. You learn a lot about the market you've chosen to go after. You learn about the competition, and a whole lot more.

Then you have to decide whether to you want to see "the flop," that is the next year to 18 months. The price to see that is usually higher. If you don't like your cards (i.e., your management team, your market, the competitive dynamic, etc.) then you fold. Cut your losses. Preserve your capital. Wait for the next deal.

In poker folding is simple. In the VC business, it's not that simple. Sometimes you can fold by selling the company or the assets. Other times, you need to shut the business down. It's not easy and many inexperienced VCs make the mistake of playing the hand out because they don't want to face the pain of folding. That's a bad move.

If you structure your deals appropriately, you can often get three or four rounds. As your hand strengthens, the cards get better, you increase the betting, putting more money at risk in each subsequent round. That's how smart poker players win and it's also how smart VCs win.

The poker analogy only works so far. Bluffing doesn't work in the VC business. If you've got a bad hand, you really can't bluff your way out of it. But on the other hand, you can impact the cards you've got. You can work with management, beef it up, switch markets, buy some businesses, etc. You can significantly improve your hand if you work at it, something that's not really possible in poker.

That's why I think VC is mistakenly seen as risky. Sure the ante is very risky. But if you play your hand right, the subsequent rounds are much less so, and the fact that you can put most of your capital to work in the later rounds makes the total portfolio a much less risky proposition than the upfront ante.

Contributed by Fred Wilson of Union Square Ventures: http://aVc.blogs.com

some of them will do phenomenally well. To produce high returns, venture capitalists must take a large amount of equity and frequently demand board representation in a developing company in exchange for their investment.

Aside from the institutional support that endorsement from a top-flight venture capital group provides, venture capitalists also lend their experience and expertise in corporate development. While a seasoned entrepreneur may have participated in perhaps a few initial public offerings, buy-outs, or other liquidity events, a venture capitalist will likely have experience with over a dozen such events. In guiding companies through early development, venture capitalists also develop relationships with other business professionals that can be invaluable to young companies.

Venture capital backing can also provide greater opportunities for further financing (including access to underwriters for an initial public offering), or lucrative corporate partnerships that can lend additional support. The ability of venture capitalists to tap networks of scientific and business professionals can aid in recruiting employees, executives, directors, customers, other investors, and influential business service firms.

MERCHANT BANKS

Merchant banks engage in many of the same equity funding transactions as venture capitalists. In some cases the distinction between venture capital and merchant banks is blurred because of the similarity of their activities, but important distinguishing elements lie in the funding and operational structures of the two. Venture capitalists typically raise objective-driven funds from a defined group of limited partners, and investments are managed by a small group of venture partners. Merchant banks display far greater flexibility in their structure and in their investments. Additionally, merchant banks tend to be far less directly involved in the operational activities of their investments than venture capitalists, and often favor buy-out exits.

INCUBATORS

Incubators provide workspaces and business services to promising companies in exchange for equity. Unlike industries in which nascent companies can use a garage or cheap warehouse space as their first workspace, biotechnology requires facilities designed to handle dangerous chemicals and perform sensitive experiments (see *Facilities* in Chapter 10). Incubators can provide lab space at a discounted price, allowing companies to mature to the point of self-sufficiency. Business services are often also included, ranging from secretaries and accountants to business training sessions. Ideally, incubators are formed with the mission of encouraging interaction among, and providing business assistance to, developing companies. Formal programs to coordinate and deliver business assistance are common, and the ultimate goal of an incubator should be to support the development of self-sufficient companies.

The development-by-incubator model got a bad reputation during the Internet boom in the late 1990s when many so-called incubators were able to draw exorbitant amounts of equity from multiple ventures and passively wait until some were fortunate enough to succeed. In biotechnology, lengthy commercialization times and the great expense of research and development discourage uncommitted incubation-for-equity projects. Instead, many biotechnology incubators are sponsored by local universities or regional governments with the intention of helping develop university inventions or expanding the local economy.

EXIT

Entrepreneurs and financiers may cooperate in the development of a venture, but they have dissimilar goals. While entrepreneurs may desire to develop or commercialize a product or service, financiers are generally primarily motivated to gain a return on their investment. This does not imply that descriptions of financiers as money-hungry are always justified, but financiers must be permitted to profit from investing in a company. Investors seek to purchase equity in promising companies that reflects the risk of failure. They accept the risk that they will not recoup their full investment in exchange for the possibility that they will turn an appropriately sized profit.

Even if entrepreneurs and financiers are both primarily interested in financial returns, financiers usually have a liquidation preference, enabling them to obtain the first proceeds from any sale of the company. This liquidation preference is necessary to marry interests with the entrepreneur who, having invested less capital than the financiers, could otherwise simply sell their shares at a profit without building any value for the financiers.

Two common exits for investors are initial public offerings (IPOs; described later in this chapter) and sale of private companies through merger or acquisition (described in Chapter 14). While IPOs are often favored by founders, as they enable a company to maintain its independence, acquisitions may offer greater returns. An analysis of biotechnology transactions from 2003-2005 found that median return for IPOs was 2x, while acquisitions delivered a median return of 3.5x.[2] While the choice to pursue a merger/acquisition or an IPO is complex and influenced by current market dynamics, a company's business model, and other elements, this observation does illustrate that IPOs are not necessarily the preferred option.

It is important to consider the timelines of investors. While investors may be attracted to technologies and products that can have positive social and scientific outcomes, they must also balance their investments with the projected time to exit. Investors may exert pressure for a company to grow at a faster pace, or to merge or liquidate earlier than founders or managers desire. Venture capitalists raise new funds every few years and need high-value exits in order to demonstrate their ability to deliver outstanding returns for their investors. Angel investors may have longer timelines, but it is important to consider that these investors tend to be retired executives who want to see significant progress while they are young enough to celebrate successes.

While investors may be ultimately motivated by a return on investment, they still rely on some degree of commercial success to achieve their goals. As a company succeeds in its development plans,

---

2 Behnke, N., Hültenschmidt, N. New path to profits in biotech: Taking the acquisition exit. *Journal of Commercial Biotechnology*, 2006. 13:78-85.

the risk of failure decreases and the valuation should increase accordingly (see Figure 17.2 and *Maximizing Multiples* in Chapter 17). A more certain value of a company's intellectual property, product pipeline, and developed products may motivate acquisition by larger firms or enable the sale of shares in the public markets.

PUBLIC MARKETS

Public listing is an important factor from an investor perspective because it permits the realization of gains on investments through the sale of shares granted for investments. Most investors are unwilling to wait for residual revenues from cashflow, and seek liquidity opportunities such as selling their equity shares in public markets or to an acquiring firm.

Companies sell shares in public markets to access greater sums of money than would be available through private markets, as well as far greater dissemination of shares among investors, which typically leads to fewer restrictions on the company, without a control block or lead investor. In raising money, private markets operate very much like public markets; investors provide cash in exchange for shares. Fluctuations in private and public market conditions can affect the number of shares that investors expect to receive for a given investment. A fundamental difference between public and private markets is that public markets are more liquid—it is easier to sell publicly-traded shares than privately-issued shares (see the requirements to be an accredited investor in private equity in the discussion of *Angel Investors* above). The relative inability for investors to divest themselves of investments in private firms leads them to expect a price discount. Consequently, public firms tend to carry higher valuations than similarly positioned private firms. The ease of investment in public companies also enables investors to readily invest in undervalued firms, permitting rapid recovery of share prices.

Public registration also offers many advantages for companies. The greatly increased liquidity (and objective value) of publicly traded shares relative to privately traded shares facilitates their use *in lieu* of cash. Shares, rather than cash, can be used to form partnerships or to raise equity, and possibly acquire, other companies. Additionally, granting time-restricted stock purchase options to employees and constituents can also reduce cash expenditures and improve retention. Public registration can also grant credibility with investors, since public companies have increased reporting requirements and must file audited financial statements, and can provide an objective measure of a company's valuation. Because investors in private companies often anticipate public offerings as a means to realize gains on their investments, the potential of a public offering can attract funding while a company is private.

With the financial advantages of public registration come additional responsibilities and ever-changing regulations. While private companies have few requirements to disclose information, public companies must report any event that materially affects profitability. The requirement to report significant scientific findings and failures may give competitors an advantage. Furthermore, while most day-to-day scientific events will not significantly affect a large company's profitability, the same is not true for smaller public companies. The requirement to consistently show strong performance to sustain stock prices, a challenge for a biotechnology firm of any size, can also be taxing on senior management.

Internal control is sacrificed in the transition from private to public ownership. When markets turn against biotechnology, it can be difficult for public companies to raise funds for their operations. Employees facing a precipitous drop in the value of their stock options may be motivated to find alternate employment, disrupting operations and growth. Additionally, poor performance may lead disillusioned investors to file suit against management or demand management changes (see Box *Shareholder activism* in Chapter 10). A recent evaluation of 100 public biotechnology firms found 31 of them facing lawsuits, 21 of which had recently faced a drug development setback.[3]

Public companies also face the risk of unwanted acquisition. Should the market value of a company fail to reflect its perceived value, an unwelcome suitor may initiate a hostile takeover, purchase a controlling interest in the firm, or initiate a proxy fight to unseat

---

3 Travers, C. Drug fails, get sued. *Motley Fool*, July 19th 2004. http://www.fool.com/news/commentary/2004/commentary04071901.htm.

Biotechnology Funding Sources

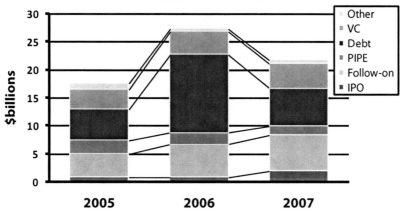

Figure 11.2 *Biotechnology funding sources*
Source: Burrill & Company

the existing board of directors (see *Shareholder activism* in Chapter 10).

INITIAL PUBLIC OFFERING

An initial public offering (IPO) is the first sale of stock by a private company to the public, marking the transition from being privately traded to publicly traded. Unlike the exchange of shares in the open market following the IPO, cash proceeds from the initial offering (less underwriter discounts and expenses) goes to the listing company, providing an opportunity to raise significant amounts of capital.

An IPO requires interest on the part of the public to purchase stock in the newly listed firm. The concept of IPO windows is central in engaging in an IPO. The window is considered open in a given sector when markets are receptive to IPOs. Failure to obtain sufficient interest in the "roadshow" or marketing of the IPO by the company and its underwriters may result in cancellation, or postponement with a potentially decreased offering price and size. Preparation for an IPO can be lengthy and expensive, and may distract management from other responsibilities, so careful timing is required to ensure that preparatory activities can be performed in advance of window opening, and before it closes.

REVERSE MERGER

Initial public offerings are not the only way for a company to become listed for trading in the public markets. Just as acquisition by a public company can effectively make a private company public, a private company can gain public listing by merging into a public "shell" company, which allows the private company to immediately become public. A shell may be a public company with few or no assets that may be the remnant of a bankruptcy or asset sale, or which may have formed as a special purpose acquisition company (SPAC) with the intention of engaging in a reverse merger. The quality of a shell corporation is critical, as merging with a shell carrying debt or pending lawsuits related to prior business activities can be very detrimental.

Some of the benefits of reverse mergers over IPOs are that they are often faster and less expensive than IPOs, and they may enable a company to gain public listing at times when public markets are not

---

Box

Alternative route to going public

More and more small and medium-sized investment banking firms are devoted to raising equity capital for private, emerging growth companies including biotechnology companies as they become publicly held through the reverse merger process.

For most private growth companies, an initial public offering may be infeasible, too expensive or time-consuming, or lack sufficient investment banking interest and market reception. For these companies, a reverse merger can offer an attractive alternate route to raising growth capital and becoming public. In such a transaction, the private company is merged with a publicly traded "shell" company with current SEC filings, and immediately becomes public as a result. This process provides several advantages to the company, including:

- A direct pathway to liquidity for the company's

shareholders
- A currency for strategic acquisitions and employee retention
- Greater access to the capital markets

Biotechnology companies, in particular, may need successive infusions of capital during the product development cycle. Once publicly traded, biotechnology companies can potentially expand the range of investors available and willing to finance them. Executives of private growth companies may find that these advantages outweigh the drawbacks of being public, such as compliance with the Sarbanes-Oxley Act of 2002. If thoughtfully structured and well executed, a reverse merger can significantly enhance shareholder value for the private company.

In conjunction with the reverse merger process, the investment banking firm would serve as a placement agent for private equity capital, in the range of anywhere from $5 million to $25 million. From an issuer's perspective, coupling a round of financing with a reverse merger can significantly lower the cost of capital. Frequently, this financing technique compares more favorably with the valuations and attendant dilution presented by traditional venture capital firms, primarily because of the premium associated with a public market for a security.

Many institutional investors, including hedge funds, arbitrageurs, small cap equity advisors, distressed securities funds and proprietary accounts for major Wall Street brokerage firms, have been regular sources of capital for issuers in these transactions. In general, these sophisticated investors have broad investment parameters and are seeking high risk-adjusted returns, on a diversified basis, among various alternative asset classes.

Once the reverse merger and financing have been completed, the difficult tasks of creating a fair aftermarket valuation and liquidity in the company's stock begin. A financial public relations program that includes investor meetings and conference calls (and, potentially, independent equity research coverage) is key to raising the profile of the company's stock and thus enhancing its liquidity and, perhaps, its trading price.

Contributed by Spencer G. Feldman of Greenberg Traurig, LLP: http://www.gtlaw.com

favorable to IPOs. Benefits of IPOs are that they can raise significant amounts of cash, and they can also create a high profile for the company, which is beneficial for investor awareness. Just as liquidation preference protects investors by preventing founders from selling their shares without adding value, the principals of reverse mergers and IPOs are typically restricted from selling their shares for 6 to 12 months after the transaction to protect the investments of public shareholders.

SECONDARY OFFERING

A secondary offering is the public or private sale of stock subsequent to a company's initial public offering. Genentech engaged in two secondary public offerings following their second IPO in 1999, raising nearly $6 billion (see Box *Genentech: Commercializing a new technology* in Chapter 2).

Secondary offerings may be dilutive or nondilutive. Nondilutive offerings involve the sale of existing shares, such as founder, private equity, or other existing shareholders selling their shares. Dilutive offerings involve the company issuing additional shares, with approval of its board of directors, for the purpose of the secondary offering. The issuance of additional shares dilutes the percentage ownership of existing shareholders, and provides the company, rather than existing shareholders, with cash. While dilutive offerings have a negative short-term impact for shareholders, the long-term benefits of providing additional cash for the company should ultimately benefit investors.

Convertible offerings enable a company and its investors to benefit from future stock price increases while being shielded from unexpected negative performance. Preferred stock, an example of a convertible offering (along with convertible and debt securities), cannot be sold and is purchased with the stipulation that it will convert to common, sellable, stock at the option of the holder. Preferred stock is often offered at a discount to the common stock price at the time of offering, providing a shield against negative performance. This arrangement enables investors to obtain large blocks of shares at discounted rates while a stock-offering company is assured that investors will not immediately sell their shares and drive down stock prices.

PIPE

A PIPE (private investment in public equity) is a form of equity financing that enables private investors to purchase shares of an already publicly-traded company, often at a discount to the prevailing market price. Requiring the use of SEC registration rights or Rule 144 stock holding-period provisions prevents immediate sale of shares, making them similar to convertible offerings, with some important regulatory distinctions involving the conversion to liquid stock.

There are two essential differences between PIPEs and buying shares in public markets. First, money paid for shares issued for a PIPE goes directly to the company rather than to investors selling shares, enabling a company to benefit directly. Second, investors can realize a discount to market rates and avoid liquidity problems associated with large transactions. Like other investor groups, institutional PIPE investor funds have been established to participate in this market.

OTHER FUNDING SOURCES

Private companies are restricted in the ways in which they can issue shares, and the lack of liquidity of private shares devalues them in comparison to publicly-traded shares. Many private companies are also too small, or have too much uncertainty in their future prospects, to attract equity investments. These companies must look to alternate sources of funding.

When the outlook for biotechnology companies is positive, companies can raise significant amounts of money, enabling them to fund operations and foster growth. When public markets turn against biotechnology, public and private companies can find themselves challenged to raise funds. Public firms may find that trading equity for cash when markets are down will overly dilute company ownership and hurt existing shareholders. Weak public markets can also affect private companies, as investors may lower their exit expectations and demand a relatively greater share of equity in exchange

Box

Focus on free cash flow

Free cash flow is the essence and life-blood of any organization, particularly a small growing company. The common definition of free cash flow is the level of remaining cash available to a company after all expenses, including investments, have been paid. Financial management texts define free cash flow as operating cash flow minus gross investment in operating assets, or net operating profit after taxes minus net investment in operating assets. Free cash flow provides the means for financial managers to enhance their company's value, and provide a positive cash flow statement.

Corporations that have too much free cash flow may run the risk of being cash-rich but investment-poor in terms of building the corporate infrastructure—not investing sufficiently in plant, equipment and human resources necessary for developing and sustaining growth. Conversely, insufficient free cash flow makes it difficult to cover necessary investment in the corporate infrastructure.

However, this balance is the parody of free cash flow; a company with a negative free cash flow position is not always an indication of a poorly run company. A company that has a positive cash flow position and makes large investments in its infrastructure could produce a short-term negative free cash flow position. The important point to remember is that no matter the size of a company, the financial focus must be on establishing a positive free cash flow position. Sound financial management practices which yield strong earnings per share are a good indication of a well managed company. However, it is the level of free cash flow which signifies the financial viability of a company of any size.

For the small, emerging biotechnology company, these are essential for sustaining business operations, establishing solid growth and enhancing corporate value. A biotechnology company initially funded by SBIR/STTR grants must look to create immediate cash flow through viable and readily available revenue streams such as sale or licensing of its technology or the sale of therapeutics and treatment practices in order to create free cash flow. As a start-up transitions to the small growth company status, fundamental activities such as understanding market demand, identifying the competition, assessing customer needs and producing products or services that meet customer needs should provide the revenue and the subsequent free cash flow to sustain growth.

Contributed by Gerald S. "Sandy" Graham, PMP, MBA, MS; sgraham@emdeon.com

for funding (or, relatively less funding). Acquisitions and stock offerings may also be postponed, forcing firms to scale back operations or delay expansion. Accordingly, companies need to be prepared to tap diverse funding sources.

GOVERNMENT FUNDING

In addition to the numerous sources of private funding, there are several government programs that can support biotechnology projects. Funding that supports the early stages of research that may lead to drug development is usually obtained from public granting institutions such as the National Institutes of Health, National Science Foundation, and others. Additionally, regional interests such as local and state governments and other development initiatives and private institutions can offer land grants, temporary tax and public utility waivers, loans, and cash. An advantage of government funding relative to private financing is that equity is not exchanged for funding, leaving ownership equity undiluted. The review process to receive government funding can also act as a proxy for private investors, assuring them that objective outsiders have faith in the scientific and commercial prospects of a company.

A downside of government funding is that it can distract a company from its commercial goals. For example, there are many grant opportunities that a company may be able to win but which are not directly aligned to the commercialization plan. While these grants can provide vital funding, they can also give potential investors the impression that a company's focus is on getting grants, not on developing commercial products.

This section highlights some of the more common and broad-spectrum federal funding programs. There are many other specialized federal, state, and regional programs that can also be tapped for appropriate projects.

SBIR / STTR

Implemented in 1993, the Small Business Innovation Research Program (SBIR) requires every federal agency that spends over $100 million on research and development contracts to set aside a small portion of their budget for SBIR grants to companies with fewer than 500 employees that are at least 51 percent U.S.-owned by individuals and independently operated. Seeking to "stimulate technological innovation," SBIR guidelines instruct managers to select proposals on technical merit and potential for commercial success.

The Small Business Technology Transfer (STTR) program is similar to SBIR, the most significant difference being that STTR requires researchers at universities and other research institutions to play a significant intellectual role in the conduct of each STTR project. The STTR program allows university-based researchers to join small companies, spinning-off commercially promising ideas while remaining primarily employed at their research institution. SBIR and STTR funding sources for biotechnology companies include most NIH institutes, the National Science Foundation, Department of Defense, Department of Agriculture, and Department of Energy. A list of relevant Internet sites is included in Appendix A.

There are three consecutive phases in the SBIR Program. Candidates for the later phases must have been awarded funding in the earlier phases. Phase I is a feasibility study to evaluate the scientific and technical merit of an idea. Phase II expands on the results of Phase I and aims to further pursue development. Phase III is the commercialization of the results of Phase II and requires the use of private sector or non-SBIR federal funding.

While the Small Business Authority has responsibility for monitoring and coordinating the government-wide activities of the SBIR program, the responsibility for selecting SBIR topics, evaluating proposals, and awarding funding lies with participating federal agencies. In granting awards, the applicant company's qualifications and potential are considered as well as the project's market, customers, and competition. The strength of intellectual property protection, suitability of revenue stream, and probability of securing necessary funding are also important factors. In order to succeed in the SBIR / STTR program, a biotechnology company must convert research results into an innovative, competitive technology that sells in the marketplace.

TECHNOLOGY INNOVATION PROGRAM (TIP)

The National Institute for Standards and Technology's (NIST) TIP program was signed into law in August 2007 and replaces the Advanced Technology Program (ATP), which had a successful record of supporting development of innovative technologies in biotechnology and other high-technology research areas. According to NIST, TIP maintains the primary purpose of ATP—to help fill the gap between high-risk innovative research that promises broad public benefits and marketable products—but contains several key changes. Under TIP, the requirements for grants to have wide-ranging benefits that meet critical national needs are tightened; intellectual property rights can be vested by a university, and universities are also now allowed to lead joint ventures.

One of the motivations for governmental intervention in funding such research is to directly support research that is "too novel or spans too diverse a range of disciplines to fare well in the traditional peer-review process." TIP provides single-company grants capped at $3 million over three years, not to exceed 50 percent of total project costs. Joint-venture grants are capped at $9 million over five years, also not to exceed 50 percent of project costs. TIP may not provide funding to any business that is not a small or medium-sized business, though those businesses may still participate in a TIP funded project.

CORPORATE PARTNERSHIP

A significant source of revenue for biotechnology firms is other firms. Small firms may form alliances and partnerships with more established firms, giving them access to development, regulatory, and sales expertise that can help bring a product to market. Senior partners may offer capital and business services in exchange for product marketing rights, patent licenses, or equity in a smaller firm (see Chapter 14).

The nature of corporate partnerships can be materially different from simple equity funding. While seasoned investors can contribute experience and networks to a growing firm, their direct involvement is often limited. Corporate partners may bring actual operational assistance. Few investors have much experience with operations such as clinical trials, but corporate partners may have extensive experience in designing and managing trials, and in submitting applications to the FDA. These direct benefits may be far more valuable than cash investments.

While a corporate partnership can provide steady cash flow and reduce risk for a young company, it may also mean that much of the income from a successful product will not be realized by the start-up. This necessary evil can allow an inexperienced company to develop the expertise and broad product pipeline that will enable it to develop additional applications and potentially become independent.

As described in Chapter 14 and in the Box *Speedel: Spinning off to develop a shelved drug* in Chapter 10), alliances enable large companies to diversify research activities and leverage development and commercialization capacities without excessive commitment of capital or other resources. Common structures for alliances and partnerships are development contracts, license agreements, and equity investments.

Development contracts compensate partners for the development of a defined product. The financing partner usually retains rights to the developed product and may provide additional compensation in the form of royalties. A potential drawback of development contracts is that the developing partner may end up receiving only a slight premium over the actual cost of development in exchange for granting sales revenues to the financing partner, meaning that they will not realize the significant upside of biotechnology product development. As with excessive reliance on government grants, development contracts also have the potential to distract researchers and management from long-term goals.

Traditional product licensing compensates product developers for granting specific product rights to licensees. Typically, the licensor receives an upfront license fee, development milestones, approval milestones, and royalties on product sales. These payments fund development by the licensor and permit licensees to reward licensors for development successes. Royalties are an important factor in license agreements as they permit the licensor to realize the significant upside of biotechnology product sales.

An alternative to development contracts and licensing is for senior partners to make equity investments in start-ups in return for certain rights. This method grants start-ups capital and enables senior partners to share in commercial success. One potential problem with equity investments is that a senior partner holding a significant ownership stake in a start-up may be required to report a percentage of the start-up's profits or losses. Start-ups may also find that granting equity to certain senior partners may dissuade other investors and partners for fear that the start-up is effectively a subsidiary of the equity-owning partner.

It is important to align objectives on both sides of partnerships. While an emergent firm may be primarily motivated by independence, funding partners are likely to be motivated by the need for products or services. While a single large partner may fulfill all of a smaller company's external funding requirements, it may also have too much control over the junior partner's destiny. Furthermore, while large companies may make or break partnerships with minimal commercial impact, smaller firms can be devastated by the loss of a key partnership, as illustrated in the Box *Exubera: When your partner doesn't sell* in Chapter 14.

FOUNDATION SUPPORT

Many diseases have foundations dedicated to raising funds for their treatment and cure. International development foundations also sponsor research into treatments for diseases endemic to developing countries. Other foundations, like the Bill & Melinda Gates Foundation and the Archon X PRIZE for Genomics (related to the Ansari X PRIZE that awarded the first private development and launch of a spacecraft) focus on specific technical challenges. These funds can be tapped by academic researchers or growing biotechnology firms to advance research. In addition to financial support, foundations can also provide valuable networking, advice, prestige, and publicity.

Leveraging the great potential of drug sales for indications in categories such as cancer, dementia, and diabetes, some foundations take convertible equity stakes or claims to royalties in exchange for funding. If the company is successful, the equity stake and royalties can help fund other foundation activities.

A significant concern for companies reliant on foundation support is wavering support from foundations and their contributors. The long lead times for biotechnology product development allow the possibility that contributors, and by extension foundations, will lose interest in the need a company has been working to serve. Withdrawal of support can derail development efforts. Also, because certain applications, such as drugs for diseases endemic to developing countries, may rely on purchase and distribution by foundations, loss of support can leave a biotechnology company without a customer for the products it has labored to produce, potentially forcing radical alteration of its business model and commercial objectives.

DEBT

Securing debt against assets and future profits, selling convertible equity, and selling blocks of stock permit companies to raise money when markets are down, and also to finance operations until public market support returns.

While debt can be secured against physical assets such as machinery and other forms of property, a preferable form of debt is to exchange future profits for immediate cash. For example, a biotechnology company seeking funds to commercialize a product may accept a cash investment in exchange for future royalties. Drawbacks of debt financing are dependence upon financier confidence in the ability of a company to deliver royalties and the possibility of seizure of collateral that is vital for success. Accordingly, debt financing tends to be restricted to companies near commercial launch.

Examples of debt financing used by biotechnology companies include using intellectual property such as patents covering developed drugs as collateral (see Box *Merck's Bermuda subsidiary runs afoul of U.S. tax laws* in Chapter 16), and selling royalty streams for immediate cash. A downside of using IP as collateral is that the lender may not fully value the IP in determining interest rates, and defaulting on the loan may result on loss of the IP. Selling future royalty streams for immediate cash likewise discounts the value of the royalty stream.

VALUATION

Determining the value of a biotechnology company is a key element of financing. Because biotechnology product development is an especially expensive and risky endeavor, it is not practical—and seldom possible—to obtain all the funding required for development at the outset. Instead, emerging companies raise funds in stages.

Under ideal circumstances, a biotechnology company will trade decreasing proportions of equity for increasing proportions of funding as scientific concepts are developed into viable products (see Table 11.1). The basis for this progression is the decreased risk of commercial failure or underperformance (and accordingly greater certainty of return on investment) that ensues as research is conducted and funds are consumed.

A key challenge in agreeing on the amount of equity to exchange for funding is determining the value of the company in question, which often has few tangible assets and minimal cash flow. Three common methods used to determine valuations are discounted cash-flow (DCF), options pricing, and comparables. These three methods use different assumptions and models to derive their valuations, and vary in relevance depending on the context of a valuation. Most valuations ultimately employ more than one of these methods.

DISCOUNTED CASH FLOW

Discounted cash flow analysis determines the present value of a product or company by calculating anticipated cash flow and discounting this future value by applying a discount rate. The result of this analysis is a net present value (NPV) or risk-adjusted net present value (rNPV) of the product or company, depending upon whether developmental risks have been included in the discount rate. The purpose of applying the discount rate is to account for the time required to reach the projected cashflows, and the risk of not being able to achieve them.

Consider a simplified example of a company with a single product in development—a potential best-in-class cancer drug with anticipated annual sales of $1 billion in 10 years. After considering the potential lifetime sales of such a drug and subtracting the costs of development, marketing, production, distribution, and sales, the future profits must still be discounted by a factor that considers the risk of failing to successfully develop the drug and the cost to investors of having their cash utilized by the start-up and not being able to invest in other opportunities for the initial ten years.

Disagreements between investors and investment recipients are common for every element of DCF: projected market sizes, risk factors, cost of development, etc. Although they may appear to be very methodical and objective, DCFs are highly subjective. Market size projections from different parties can vary by an order of magnitude, and determining discount factors involves a great deal of subjectivity (see Box *Remicade: Resolving valuation disagreements* in Chapter 14). One important aspect of DCFs is that they involve a comprehensive analysis of the projected cash flows of an enterprise and the risk factors associated with achieving profits. This analysis is a valuable exercise that provides a deeper understanding of the fundamentals of a business.

Box

Valuation of biotechnology companies

Valuation of a biotech company is said to be more art than science. It certainly is a complex and difficult task, especially for early-stage, biotech companies. The so-called pre-money valuation (which takes place before a company is financed) dictates how equity is divided amongst a company's investors and entrepreneurs.

A company's value lies in its potential to generate a stream of profits in the future. Profits can be generated from sales of drugs, services but also from up-front, milestone and royalty payments. All valuation exercises are based on "visioning" a company's future, relying almost entirely on educated guesses. To generate these assumptions feedback must be solicited on three factors: first, the state of the market targeted; second, the company's science and technology; and third, the ability of management to deliver on the business plan.

There is no "silver bullet" when it comes to valuation: it remains a subjective task. Nevertheless, it is recommended that every valuation start with a systematic and rigorous testing of a company's economic, technological, and managerial hypotheses in combination with the following two key approaches:

- Primary valuation, which is based on such fundamental information as projected future free cash flow (FCF) and costs of capital. Decision tree analysis is often used with probabilities for the success of each clinical phase and estimates for development cost, up-front and milestone payments, royalties, costs and sales of the product on the market as well as the form collaboration with licensing partner.
- Secondary valuation, which is based on comparable information, where valuation is done by analogy to other similar companies. For public companies value information is available from the stock exchange. For private companies, information of the last financing rounds can be utilized.

The various valuation approaches are likely to provide different figures, since they all capture different drivers of value in the firm. Reconciling these figures requires an extensive understanding of these drivers. Finally, although the valuation methods described here are routinely used by investors, we offer three important cautionary remarks to help the newcomer to watch for typical pitfalls:

1. Pay attention to pre- and post-money valuation; this can make a big difference
2. Know your figures before you enter negotiations
3. There is more to a deal than valuation

Contributed by Dr. Patrik Frei of Venture Valuation: http://www.venturevaluation.com

OPTIONS PRICING

While discounted cash flow analysis attempts to consider all future events and determine a discrete present value based upon the variety of possible outcomes, options pricing employs an incremental analysis. By staging investments and valuations on key events and outcomes, investments can be made slowly as more is learned about the value of the investment. A company would not, for example, seek FDA approval for a drug that failed clinical trials. The milestone-based investments made in biotechnology companies are essentially a form of options pricing. Investment in each developmental or clinical trial phase grants an option to invest in the next.

Options pricing, like DCF, is a valuable exercise because it requires a comprehensive analysis of the elements necessary for commercial success. A downside of options pricing is that it is more complex than DCF and, while it yields more precision than DCF, it is not necessarily more accurate.

It is also important to develop an understanding of the unique risks involved in each case when using methods involving risk-adjustment. Biotechnology product development is very risky, and the costs and risks of individual products vary greatly. Estimates based on discrete discount factors should account for this variability, which is why human judgement is so often used in making R&D decisions (see Figure 15.3 in Chapter 15).

COMPARABLES

Comparable valuations are derived by comparing valuations of similar companies and products. Criteria used to compare companies and products may be metrics such as the number of drugs (or other products) in development, therapeutic (or other application) categories, and progress in clinical trial phases. Examining recent company valuations, as defined by IPO performance, purchase price in a merger, acquisition, or private equity investment can be used to develop a set of comparables. Products may likewise be valued by examining revenues of comparable marketed products (and potentially discounting if assessing the valuation of a product in development), deal terms in acquisition or licensing agreements, or terms of partnership or development alliances for similarly-positioned products.

Most valuation exercises ultimately use comparables to ensure that valuations are in check with recent transactions. One significant issue in comparable valuations is the influence of market sentiment. Major setbacks can cause the value of all companies in a category to fall, regardless of merit, and irrational exuberance can likewise illegitimately inflate the value of companies in a category.

Chapter 12
Research and Development

I am not discouraged, because every wrong attempt discarded
is another step forward.
Thomas Edison

Because biotechnology is focused on commercializing innovation, research and development are central to any company's strategy. R&D is central in biotechnology because it is the means by which companies develop innovative products which are worthy of patent protection and which can sell for premium prices. While some companies may eschew research and focus instead on developing or commercializing leads produced by others, R&D is still critical to their success—even if they are not directly involved. This chapter focuses on the process and strategies of R&D.

As described in later chapters on marketing and management, R&D is not the endpoint for biotechnology companies. Developed products must address meaningful needs of customers who are able to pay profit-enabling prices. R&D management is discussed in Chapter 15 along with other biotechnology management topics.

NON-DRUG BIOTECHNOLOGY

An important distinction between drug development and other biotechnology applications is that drugs are highly regulated—during development, and after they are in the market. While this high degree of regulation can be a burden to drug developers, it can also provide some substantial benefits. As described in the Box *Clinical trials provide valuation milestones*, and depicted in Figure 12.1, the

> **Box**
> # Clinical trials provide valuation milestones
>
> The process of taking a novel idea and developing it into a marketable biotechnology product is very long and expensive. Funding for biotechnology products is rarely provided as a lump sum. Instead, funding events are often attached to key developmental achievements called milestones.
>
> Clinical trials represent an objective measure of the progress of a drug from a research lead to a marketable product, giving drug developers a discrete set of measures to which funding can be aligned. Non-drug applications, such as agricultural or industrial biotechnology, lack these objective measures, requiring the use of substitutes such as *prototype* or *proof-of-concept* which are subject to interpretation (see Figure 12.1). By aligning financing events with clinical trial progress, or by citing clinical trial progress in seeking new investors, these objective measures can be used as a metric for development progress.
>
> In addition to providing objective third-party criteria by which the development status of drugs can be measured, clinical trial phases also create a marketplace for drugs in development. The likelihood of approval for a drug in clinical trials can be projected by comparing historical approval rates of similar drugs. Weighing this factor against the estimated market size for a drug can help assign a financial value for drugs in development (see *Valuation* in Chapter 11) which, in turn, facilitates the sale of drugs in development. Selling drugs in development can enable a company to realize the value of drugs it lacks the interest or resources to commercialize, and can fund other research projects.

third-party measures of clinical progress and therapeutic indications for which a drug is effective, provided by regulatory authorities, can be leveraged to facilitate financing or sale of drug leads.

R&D STAGES

Drug discovery is described in greater detail in Chapter 4, and the general scheme of R&D is summarized here. There are two fundamentally different kinds of research: basic and applied research.

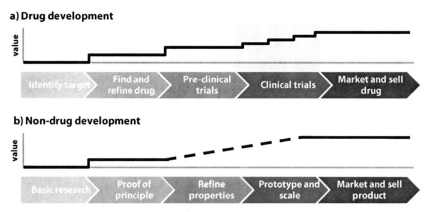

Figure 12.1 *Clinical trials provide valuation milestones*

Basic research is directed at improving fundamental knowledge, whereas applied research is directed at applying knowledge gained from basic research. Different toolsets, mindsets, and players are involved in these types of research. Basic research does not directly produce biotechnology products. Instead, it lays the foundations upon which products are developed. Applied research is required to enable further development.

The basic stages for drug and non-drug development are shown in Figure 12.2. The stages are segregated by fundamental differences in operations and objectives. For example, different mindsets, skill sets, and deliverables are involved when discovering a potential drug versus refining the properties of a drug lead.

While the stages are presented here as a discrete linear process, in practice the distinctions between stages are not so distinct. Promising leads may be advanced to development while discovery-stage research continues to investigate basic-science questions and additional opportunities. Likewise, commercialization activities need to be initiated while discovery-stage research is occurring. Waiting for FDA approval before establishing manufacturing capacity or initiating reimbursement discussions with healthcare payers is not practical due to the amount of time required to prepare for these activities.

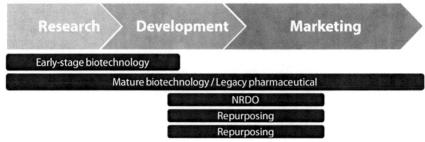

Figure 12.2 *R&D Stages*

DISCOVERY

Early stage R&D focuses largely on defining the system being examined and identifying lead compounds. At this stage the primary objective may be to simply study a disease mechanism in the search for druggable opportunities, or to identify compounds that interact with disease mechanisms. The high rate of attrition of leads emerging from discovery-stage research is a motivation for many companies to license leads from others, or to form research alliances. *Outsourcing Innovation*, later in this chapter, describes some other methods used by established firms to expand their access to novel research.

DEVELOPMENT

Once leads have been identified and have passed early tests to demonstrate promise, it is necessary to ask a new set of questions. Development-stage researchers strive to refine lead compounds, seeking to expand the understanding of how they work, to investigate opportunities to improve stability and efficacy, and to reduce potential side effects. The development stage is also where pre-clinical and clinical trials occur (see *The Five Basic Steps of Drug Development* in Chapter 4) to determine the safety and efficacy of the drug in humans.

During the development process the market potential of a biotechnology product also needs to be assessed. The target markets need to be identified, and the suitability of a product to serve these markets must be evaluated. The quality of a market itself must also be assessed, and potential reimbursement issues should be identified early. Marketing and reimbursement are discussed in greater detail in Chapter 13.

COMMERCIALIZATION

Once a drug has received FDA approval (or, in the case of other biotechnology applications, the product has completed development) it is time to scale-up production and start selling and distributing the product. As mentioned above, preparations for commercialization need to be initiated well in advance. Limited patent life and competition encourage companies to move quickly once products are ready for sale.

Commercialization activities also extend beyond simple manufacturing, marketing, and selling of products. Once a product with established safety and efficacy is on the market, it is time to start leveraging those attributes. Repurposing, described under *R&D Strategies* below, seeks to find novel applications of established drugs. A drug that is safe for one indication is likely to be safe for other indications: the challenge is to find additional applications where it is effective. Reformulation strategies, also described below, can extend a product's life after patents have expired by enabling the filing of new patents on improvements in delivery systems or efficacy.

R&D STRATEGIES

The essence of research and development is that it is the process by which commercial ideas are "de-risked" through testing hypothesis, and developed into commercial products (see *Reducing Risk and Making Money* and Figure 17.2 in Chapter 17). Drugs are the prototypical application of biotechnology companies, but R&D is also necessary for applications in other areas such as research tools, agricultural, or industrial biotechnology. Selected R&D strategies are discussed in this section. A general overview of biotechnology business models is presented in Chapter 10.

REPURPOSING

Repurposing is the search for new purposes for existing drugs, generally performed by a company other than the originator of the

---

Box

The cost of drug development

When assessing the cost of drug development, most sources cite one of two studies from the Tufts Center for the Study of Drug Development (CSDD). In 2003 the CSDD estimated the average cost of developing an approved traditional pharmaceutical drug to be $802 million, and measured the development time to be 10-15 years.[1] In 2007 they estimated the cost of biologic drug development to be $1.2 billion, with a development time greater than 12 years.[2] It is worth noting that the biologic cost estimate derives from a relatively small sample size: 17 investigational drugs from four firms. These estimates are valuable because they measure the overall cost of drug development programs, but it is important not to use them as metrics for the cost of developing individual drugs.

These estimates do not suggest that a company would need to invest $1.2 billion, or even $800 million to develop a new drug. First, the estimates include the opportunity-cost of capital, which is an expression of the lost income opportunities because investments must be made years in advance of approval—the opportunity cost of not being able to invest money in stocks, bonds, etc. Actual cash outlays for biologic and small molecule drugs were respectively estimated to be $500 million and $400 million. Accordingly, it is not accurate to state that biologics cost $1.2 billion to develop over 12 years—the duration of drug development is already expressed in the cost estimate. Second, these estimates include development costs for failed drugs. They do not reflect the cost to develop a new drug; they measure the cost of drug development programs. The estimates were derived by assessing the cost of drug development programs and dividing that cost by the number of products ultimately approved.

Therefore, while the average time-adjusted cost of biologic development drug has been estimated to be $1.2 billion, the cash expenditures required to develop a single new biologic drug are less than $500 million.

1 DiMasi, J.A., Hansen, R.W., Grabowski, H.G. The price of innovation: New estimates of drug development costs. *Journal of Health Economics*, 2003. 22:151–185.

2 DiMasi, J.A., Grabowski, H.G. The cost of pharmaceutical R&D: Is biotech different? *Managerial and Decision Economics*, 2007. 28:469-479.

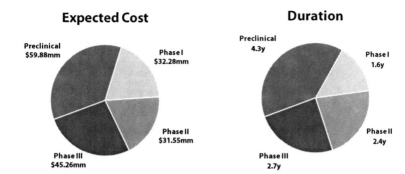

Expected Cost

Preclinical
$59.88mm

Phase I
$32.28mm

Phase II
$31.55mm

Phase III
$45.26mm

Duration

Preclinical
4.3y

Phase I
1.6y

Phase II
2.4y

Phase III
2.7y

Figure 12.3 *Estimated cost and duration of biotechnology drug development*
Source: DiMasi, J.A., Grabowski, H.G. The cost of pharmaceutical R&D: Is biotech
different? *Managerial and Decision Economics*, 2007. 28:469-479.

drug. The same basic metabolic processes may be implicated in multiple diseases, and a single drug may also have more than one biological effect, enabling one drug to treat numerous conditions. Seeking approval for additional indications can also potentially reduce clinical trial burdens by citing known safety profiles, and may also leverage existing patent protection. Advanced biotechnology techniques like expression profiling using microarrays can help screen for additional applications of existing drugs. High-profile repurposed drugs include thalidomide (see Box *Repurposing thalidomide*), Viagra, Minoxidil, and Propecia.

There are several significant challenges in repurposing drugs for which a company does not hold the patents. To market a drug patented by another party, a company must secure a license from the patent holders. The patent holders may resist issuing a license because they are already researching new indications for the drug; they may fear the elucidation of new safety concerns which could precipitate a recall for all indications; they could fear losing face if sales for the repurposed indication exceed those of the original indication; and, they could simply lack sufficient incentive to issue a license. If the repurposed indication is too successful—if an outside party can generate significant revenues from an overlooked indication—someone at the originating firm may face criticism or even termination for failing to recognize the potential market opportunity. It is far less likely that someone will lose their job for *not* licensing a drug.

Challenges for repurposing drugs lacking patent protection involve protecting the market for the new indication. Without patents to prevent others from producing generic versions at costs reflecting their lack of research and clinical trial costs, how can a firm recover the costs of repurposing? One option is to focus on indications eli-

---

Box

Repurposing thalidomide

Thalidomide was originally developed as a treatment for morning sickness in pregnant mothers. The drug was never approved in the United States, but in the late 1950s and early 1960s it was sold in 46 countries as a sleep aid and treatment for morning sickness in pregnant women. Thalidomide was recalled following the discovery that it caused severe birth defects. Although it is a small-molecule drug, not a biologic, thalidomide provides a case study for the potential of repurposing.

The cause of thalidomide-related birth defects lies in the chemical structure of the molecule. Thalidomide exists in two different chemical forms which are mirror-images of each other, (R)-thalidomide and (S)-thalidomide. They have identical chemical compositions but different shapes, and readily convert between forms. The (R) variant is effective against morning sickness, but the (S) form causes birth defects.

In further studies on the biological activities of thalidomide, it was discovered that the drug had applications beyond morning sickness. Based on the observation that U.S. AIDS patients were illegally importing thalidomide to treat wasting, Celgene, which started as a spin-off from a chemical company, sought to seek approval of the drug for this indication. The rationale was that despite the drug's tragic history, the FDA would likely approve the drug for AIDS patients.

Facing setbacks in demonstrating efficacy for HIV treatment, Celgene received FDA approval in 1998 to market thalidomide for leprosy under the brand name Thalomid, and later gained approval for multiple myeloma. The FDA has applied special restrictions for physicians, pharmacists, and patients (who must comply with contraceptive measures) to control off-label use and prevent birth defects. Thalidomid sales, which totalled $447 million in 2007, helped Celgene reach profitability in 2003, 17 years after its formation.

gible for orphan drug status or other regulatory-based exclusivity (see *Market Exclusivity* in Chapter 7).

This challenge of protecting markets also exists when licensing drug patents from others. While licenses may prevent patent holders from actively marketing their drug for the repurposed indication, there is little to prevent physicians from leveraging cost differences and issuing off-label prescriptions of either the patent-holder's or re-purposer's drug for the other's indication (for an example of con-trolling off-label use, see the Box *Avastin: Controlling off-label use* in Chapter 13). Solutions include aligning prices (which may limit the sales potential or profitability for the repurposer) and restrict-ing repurposing to formulations and dosage methods for which the repurposed drug cannot substitute for the original.

COMBINATION AND REFORMULATION

Combination and reformulation strategies can improve the ther-apeutic properties of drugs. Because novel combinations and refor-mulations may be patentable, these strategies are also vital tools in extending patent-protected life. Because these strategies overlap— some reformulations are combination products—they are described together.

COMBINATION

As products mature, the motivation to combine them increases. After a drug has been on market for several years, the developer will have extensive knowledge on safety and efficacy well beyond that collected in initial clinical trials. Developing combination products provides an opportunity for biotechnology companies to leverage this knowledge and potentially extend patent life by developing new and useful drug combinations.

Combination drugs have existed in a relatively simple form for decades. Combinations of antibiotics and HIV cocktails have pro-vided multifaceted approaches to disease treatment. These work by targeting numerous disease processes simultaneously, and are gen-erally intended to prevent infectious disease. Advanced combination therapies integrate elements with very different modes of action.

Drug coated stents, for example, combine stents which physically prevent artery narrowing with drugs that prevent reblockage.

A significant challenge in developing combination products is optimizing the delivery method and preventing interactions between the combined drugs. Delivery for a single drug—biologic or small-molecule—is not a trivial task, so ensuring that multiple drugs can be combined into one delivery system and achieve the appropriate blood concentrations for efficacy can require significant R&D.

Beyond developmental issues, it is also important that combination technologies and products be protectable. As with traditional drug development, innovators require patent or other protections to prevent competitors from leveraging the R&D efforts of innovators and selling competing products at a cost reflecting the reduced development costs. As described in the section *Non-obviousness* and elsewhere in Chapter 7, opportunities for patent protection include demonstrating unexpected benefits of combining products, and novel processes in the assembly or delivery of combination products.

REFORMULATION

Altering a drug's formulation or delivery method can yield improvements in safety, efficacy, and ultimately result in preferential prescription and increased patient compliance. Just as repurposing can leverage existing safety data, reformulation can also significantly reduce the cost of developing improved drugs.

A pioneer in reformulation is Enzon, the fifth company to gain

---

Box

Inhalable insulin: Holy grail or tiger trap?

The therapeutic insulin market is a mature market. Diabetics have been taking insulin for decades and many companies produce insulin products. Insulin treatment of diabetes requires regular drug administration, frequently by injection or surgically-implanted pumps, creating an opportunity to innovate and gain market share by reducing the burden or discomfort of drug dosage.

Sanofi-Aventis introduced a once-daily long-acting insulin

product named Lantus in 2000, which resolves some disease is-sues, but must be complemented by a fast-acting insulin after meals. The long-awaited breakthrough for insulin treatment, however, is a drug that does not require injections, pens, or surgical implants— an inhalable insulin product. In 2006, Pfizer launched Exubera, an inhalable insulin utilizing Nektar Therapeutics' dry powder aerosol technology. Although the primary innovation in Exubera was in the delivery method, not in the drug compound itself, clinical trials were not easy. Phase III trials were initiated in 1999, and proceeded for six more years before approval was granted.

Originally hailed as the most significant innovation in diabetes treatment in decades, and despite sales projections in the billions of dollars, Exubera was pulled by Pfizer in 2007, who cited long-term safety concerns, the need for extensive patient training, and patient reluctance to use the cumbersome and conspicuous inhala-tion device. Pfizer took a $2.8 billion pre-tax loss (approximately $1.1 billion of intangible assets, $661 million of inventory, $454 million of fixed assets, and $584 million of other exit costs) after posting only $12 million in Exubera sales in 2007, but the fallout is much greater. Nektar Therapeutics, who developed the inhalation tech-nology, was only informed of Pfizer's withdrawal indirectly by press release, calling into question Pfizer's status as a "partner of choice." The complementary Box *Exubera: When your partner doesn't sell* in Chapter 14 describes the impact on Nektar.

Unresolved issues about delivery of drugs through the lungs also remain. Lungs are designed for gas exchange, not as vehicles to deliver proteins to the bloodstream. As witnessed in gene thera-py trials, human biology can be very complex. Human biology has developed many mechanisms to repel external substances, and the potential for long-term side effects or immunological resistance to inhaled drugs represents a significant liability for any company pur-suing inhalable drug delivery. Data from Pfizer appear to bear out these concerns: there were six cases of lung cancer (among smok-ers) in 4,740 Exubera-treated patients, versus one control group of 4,292 patients.[1]

Shortly after Pfizer's market withdrawal of Exubera many other companies shelved their inhalable insulin projects as well, calling the future of inhalable biologics into question.

1 Kling, J. Inhaled insulin's last gasp. *Nature biotechnology*, 2008. 26:479-480.

approval for a biotechnology-based drug. Enzon's Adagen, which gained approval in 1990, is a reformulation of adenosine deaminase enzyme (ADA), a drug for severe combined immunodeficiency disorder, with polyethylene glycol (PEG). PEG, which is used as a thickener and foam stabilizer in food and cosmetic products, can decrease immune-system reactivity and increase the blood circulatory time of drugs. Without PEG, ADA is ineffective.

A significant benefit of reformulation is that alterations may be patentable, enabling a company to secure rights to an improved version of a drug which lacks other patent protection. This strategy is commonly used to extend the lifespan of drug brands after the initial set of patents expire. The scope of patent protection in such a case is limited to the advanced formulation. Competitors are free to sell generic versions of the drug, but the holder of the formulation patent will be able to leverage their protected formulation as a marketing differentiator.

In an example of brand extension by reformulation, Pfizer was able to generate $8 billion in additional sales from 1990 to 1998 by developing Procardia XL, an extended-release hypertension drug. To develop the drug, Pfizer used Alza's OROS technology, which enables once daily dosing. Beyond simply reducing the number of daily dosages, the controlled release formulation also makes the drug more tolerable than immediate-release formulations, providing a significant marketable advantage.

In another example of reformulation, Abraxis Oncology improved on paclitaxel (the generic version of Taxol) by binding the drug to nano-scale protein particles. This reformulation provides significant patient benefits by eliminating the need for steroids or antihistamines, which are necessary to prevent hypersensitivity reactions to solvents required for other formulations.

SPECIALTY PHARMACEUTICAL / NRDO

The specialty pharmaceutical model historically referred to companies that acquired rights to drugs abandoned by large companies, usually due to insufficient revenue potential, and which then sought to gain approval for these drugs. The smaller size of the acquiring firms enabled them to profitably sell these drugs. Another variation was performing clinical trials to gain domestic approval for drugs approved in foreign markets. Specialty pharmaceutical also describes drugs targeted at specialist physicians or those requiring special handling or special disease-monitoring. Since the emergence

---

Box

Non-profit drug development

Drug development is synonymous with two financial outcomes: high cost and high revenue. These attributes stand in opposition to a non-profit business model. The great cost of drug development forces developers to focus on drugs that are likely to support sales which greatly exceed their development cost. This economic reality impedes the development of many treatments for acute conditions, conditions affecting small markets, and tropical infectious diseases that fall below the threshold of interest of drug developers.

Many promising drug leads with applications for minimally-profitable yet pressing conditions are shelved in mid-development by drug developers unable to justify the expense of completing development. By obtaining royalty-free licenses to develop these drug leads for neglected conditions, non-profit firms are able to use a NRDO model to meet these significant needs while for-profits can benefit from goodwill and tax benefits for their donations.

Leveraging the economies of scale and resources that have emerged from decades of aggressively funded drug development efforts, non-profit firms may raise money from individuals, corporations, and foundations, and utilize volunteer assistance from pharmaceutical scientists and other skilled professionals. One such case is the development of paromomycin, a treatment for visceral leishmaniasis, the second-most prevalent lethal parasitic disease after malaria.

The World Health Organization obtained rights to the injectable form of paromomycin, which had been shelved in mid-stage clinical trials, from Pharmacia (now Pfizer). Non-profit OneWorld Health then partnered with the WHO to shepherd the drug through clinical trials, and in 2005 received orphan drug approval for paromomycin from the FDA and EMEA.

of the "no research, development only" (NRDO) model in which companies license drug leads instead of developing them with internal research efforts, the two terms have merged. To avoid confusion, NRDO is used here.

The NRDO approach utilizes a quasi-virtual structure (see *Virtual Company* in Chapter 15). NRDO companies have no internal research units and may also outsource later-stage activities. Two challenges shared by virtual companies and NRDOs alike are the inability to capture tangential discoveries that emerge in the course of research, and the need for an especially strong management team.

Many important discoveries emerge in the course of unrelated research. Because research is performed by external parties, NRDO and virtual companies may not be privy to subtle observations of potentially lucrative tangents. This loss of tangential opportunities is offset by a reduced risk profile. In purchasing drug leads that meet specific development requirements, NRDO firms are able to shield themselves from much of the early-stage developmental risk inherent in drug development, although a premium price is often attached to this benefit. The lack of an internal research unit to tap for expert guidance places an additional burden on management to determine development agendas and assess opportunities.

PROJECT SELECTION

Biotechnology companies must consider the risks of development setbacks and failures, and the rewards of commercialization, in selecting research projects. Unlike industries in which successful development is relatively certain, and where meeting budgets and ensuring commercial acceptance are primary concerns, successful development is not guaranteed in biotechnology. In selecting projects, biotechnology companies must consider both the potential rewards and the likelihood of success.

Important factors for assessing the potential value of a drug in development include the size and quality of the patient population, whether the market for that particular drug is already well served, and the company's financial situation, all of which will be factors in the decision to begin clinical trials. Faced with competition from experienced firms with established development and commercialization programs, the competitive landscape may discourage small companies from developing solutions for large markets. One strategy for small companies to avoid competition with well-positioned companies is to develop solutions for markets that are too small to warrant consideration by larger competitors. Incentives such as tax credits and temporary market exclusivity awarded for orphan drug development (see *Orphan Drugs* in Chapter 8), for example, serve as additional motivations to pursue small markets.

It is important to distinguish between market size and market quality. The quality of a market is arguably more important than how innovative or complex a product or service is or how large the market is. Government incentives, the ability to avoid larger competitors, decreased price elasticity (customer insensitivity to price increases), and reduced development and marketing costs make niche markets more attractive for small biotechnology companies. In some cases drugs serving small patient populations can deliver relatively greater returns than those serving larger markets. The Box *Genzyme: Building an enterprise on orphans* in Chapter 8 illustrates how focusing on orphan drugs can yield blockbusters, despite the small number of patients in the target market.

A drug meeting a significant need for a relatively small population may warrant a higher unit price than a non-essential drug serving a larger population and may also benefit from lower marketing and distribution costs. Furthermore, a drug addressing a significant unmet need for a small patient population will likely amount to a small portion of expenditures by medical insurance firms, avoiding price pressures. Marketing a drug to a community of specialists is also likely to cost less than the large-scale marketing required for drugs with larger markets.

DIVERSIFICATION AND FOCUS

The majority of biotechnology start-ups are formed to exploit a market opportunity that is protected as intellectual property. Because start-ups thrive on exploiting disruptive changes to serve unmet needs, a start-up will thrive or fail based on the effective implemen-

Box

Dangers of not having a pipeline

For an early stage biotechnology company to be successful it needs more than a single product. Although there are many good ideas, many biotechnology opportunities are turned into companies too early, often with a single product as their sole focus. While focus is good and required, a certain percentage of time and effort needs to be spent on second generation products. In biotechnology, many great technologies fail in the transfer from research to the clinic due to toxicity issues, manufacturing scale-up, efficacy, etc. Therefore, if a company is to survive, it must have a backup strategy, and products, in the event that the lead product dies an untimely death. Most venture capitalists look for a company with a platform or at least a product that can be used for several different indications, although this can be risky if the product fails due to toxicity or other problems that affect all of the intended uses.

An example of a failure due to lack of diversification was witnessed by a small biotechnology company that was shut down after a key proof-of-concept study failed. The company had a small but very effective team who had learned many valuable lessons about product development and the regulatory process by working on a single product. This team had very effectively solved a number of key challenges in developing their product and worked well together. Despite efforts to remove risk early in the development, the key proof-of-concept study couldn't be completed until enough product had been synthesized and a formulation developed. After the failure, it was impossible to justify the burn rate of maintaining the core team and facilities. Neither the company nor the venture capitalists were able to quickly identify an alternate technology to place into the company; identifying and licensing good technologies takes a minimum of several months. It is likely that this company would have been very successful in future product development due to the lessons they learned with the first product. Unfortunately, the staff are now scattered and are more than a little frustrated at the venture capitalists who shut them down over something that was outside of their control.

Contributed by Darren Fast, Ph.D. of Solalta Advisors: www.solalta.com

tation and exploitation of competitive advantages. Aligning research efforts to exploit core competencies and intellectual property is essential to impede competitors and leverage internal assets. Failure to use a competitive advantage opens the door to competitors.

Rather than focusing on one application, a company should diversify its research activities. Products and tools with multiple uses can provide alternative opportunities should a single application fail to obtain regulatory or public approval, become a commodity, or become obsolete. Simultaneous development of multiple applications can reduce the impact of competition and development failures.

As important as diversification is, it is important that a company not pursue too many disparate objectives. One of the contributing factors to the demise of Cetus (see Box *Cetus spreads itself too thin*) was that they did not have the resources to develop any of their diverse projects. Successful biotechnology firms tend to focus their activities according to specific research areas. Different biological systems and research methods often have unique characteristics that require specialized expertise. Just as a narrow research focus can limit a company's prospects, a lack of focus can also be disabling.

Focus is also essential to attract investors and partners. If investors are unable to define a company's focus they may decide that their money is better invested in companies that they do understand, or they may interpret the apparent lack of focus as a shortcoming of management. Prospective partners may similarly shy away from unfocused companies because of failure to understand the business model or an inability to determine if an unfocused company is a competitor.

As companies mature, more diversification is generally accepted, and expected. Established companies often display much more diversification than when they were nascent start-ups. Having too narrow a focus increases the risk that opportunities will be missed. For young companies with no proven execution expertise, the risk of not capitalizing on available opportunities outweighs the risk of missing opportunities. Established firms, by definition, have demonstrated execution expertise and are often much slower to change than younger firms. These factors favor a more diversified approach to research.

Box

Cetus spreads itself too thin

Biotechnology is a great way to make a small fortune; as long as you start with a large one.
Cetus Chairman Ronald Cape, 1984

Cetus was founded in 1971 and used the $107 million raised in its 1981 initial public stock offering, the largest in U.S. history at the time, to build a biotechnology powerhouse. With top executives who had worked at major pharmaceutical firms, a talented team of scientists, and deep pockets, Cetus conducted research in the diverse areas of veterinary vaccines, diagnostic products, agribusiness applications, interferons, fructose biosynthesis, and chemicals and energy derived from plant materials. Unfortunately, Cetus did not have sufficient revenue streams to support all of its ambitious research programs. After losing $1.75 million and $2 million in 1978 and 1979, Cetus turned a small profit in 1980. Without a single major product, Cetus was spread far too thin.

Despite the 1985 development of the polymerase chain reaction procedure, a method for duplicating DNA that revolutionized molecular biology and warranted a Nobel Prize, Cetus suffered for behaving like a fully mature drug company instead of the adolescent it was. After trimming its development pipeline to concentrate on just diagnostic products, cancer therapies, and agribusiness, further inappropriate and premature allocation of resources and unexpected product failures led Cetus to sell the company to Chiron in 1991 for $700 million.[1]

---

1 Lehrman, S., Cetus: A collision course with failure. *The Scientist*, 1992. 6(2):4.

OUTSOURCING INNOVATION

Large companies tend not to be very effective at innovation. Their structure, legacy, and incentive systems tend to favor a more conservative approach to business:

- There are more small companies than large companies. Even though small companies have fewer resources than large ones, by simple probability one would expect

small companies to be responsible for a good share of
innovation.

- Small companies can afford to take greater risks than
  large companies. A company with no, or meager, profits
  has only one option: to grow. Therefore, they tend to
  take greater risks than large firms, which sometimes
  leads to greater innovative output.
- Just as small companies may face bankruptcy if they fail
  to develop innovative products, managers at these firms
  are also less likely to advance in their careers unless
  their company is successful. A company-wide necessity
  for growth often results in greater tolerance for risk
  takers. By contrast, managers in large companies who
  back risky projects which fail may face censure or
  termination.

One means by which large companies can increase their output
of new products is by licensing product leads and engaging in development partnerships with smaller firms (see Chapter 14 and Box
Speedel: Spinning off to develop a shelved drug in Chapter 10). Even
in cases where these strategies are more expensive than in-house innovation, the significant downside in terms of stock performance or
individual careers often favors transferring the risk to outside parties—effectively paying money to reduce risk.

Beyond licensing and alliances, companies can also employ directed strategies to outsource innovation. A primary driver for outsourcing innovation is that many established companies have excess
capacity to develop innovations, but cannot dedicate sufficient resources to perform early-stage research on all the possible opportunities. By outsourcing innovation to multiple partners who specialize in early-stage research, established firms can effectively reap the
benefits of broad research programs without the associated loss of
focus.

CORPORATE INCUBATION AND SATELLITE COMPANIES

Corporate incubation and support of satellite companies enable
firms to buy options in promising companies. These strategies are

best employed when a company has excess resources which can be used to support the creation of independent companies.

Corporate incubators are useful tools for established companies seeking to source new opportunities. In this model established companies offer lab space, technical and management expertise, and capital in exchange for future acquisition rights. By placing target companies in corporate incubators, with the intention of acquiring them should they succeed, the exchange of information and assistance can be far more intimate than in a simple licensing agreement or development partnership. This strategy is especially effective for founders who favor a quick and early exit. A downside of this strategy is that it may be especially difficult for incubated companies that fail to get acquired to attract other investors.

Satellite companies are similar to corporate incubation, but instead of granting lab space and expertise, the established firm grants technology licenses to emerging firms. For example, Isis Pharmaceuticals has a strong intellectual property portfolio in the field of antisense technologies, and expertise in bringing antisense products to market. Without sufficient resources to research all the potential applications covered by its patents, Isis formalized a satellite company strategy, whereby it provides a license for specific antisense applications or target markets, or licenses a drug candidate which falls outside its commercial focus. *In lieu* of charging licensing fees, Isis instead takes an equity stake in these satellite companies, enabling them to benefit from future successes without distracting internal research efforts.

CORPORATE VENTURE CAPITAL

Beyond incubating target companies and seeding them with technology licenses, another option for established firms to increase their exposure to innovations is to directly fund promising companies. These investments may be structured similarly to traditional venture capital deals (see *Private Equity* in Chapter 11), but they often also serve the greater objective of gaining exposure to new technologies. Because these investments are often directed at acquiring technologies, rather than a simple return on investment, investments may also be made in exchange for options to license or acquire technologies, rather than for equity. For example, Novartis' venture fund, the Novartis Option Fund, employs an option fee. The option fee is non-dilutive—the funded company does not trade equity for the funding—and it grants Novartis the ability to acquire target technologies later in development.

Cooperative Research and Development Agreement

Cooperative research and development agreements (CRADAs) authorize federally funded laboratories to partner with commercial firms. CRADAs can be an effective solution for start-up or virtual biotechnology companies to access skilled researchers and specialized research tools. A CRADA may be entered into by the government with a private company, university, or non-profit firm, as well as a state or local government. Under a CRADA arrangement a government laboratory may contribute people, equipment, and facilities, but no money. The collaborating party may contribute the same and may reimburse the federally funded laboratory for manpower or other resources if required. An important feature of CRADAs is that the private sector participant has a first right of refusal to license any inventions the government makes under the CRADA.

Chapter 13

Marketing

People don't buy products or services. They buy solutions to painful problems. If your customer has a headache, sell aspirin, not vitamins.

John N. Doggett, McCombs School of Business

M arketing is central to commercializing biotechnology. Although it is often thought of as a late-stage activity closely tied with advertising and sales, marketing plays an important role early in R&D. Marketing is one of the elements which distinguish biotechnology companies from traditional research labs. Biotechnology companies are not compensated by receiving grants to further productive research efforts; they are compensated by selling useful products to customers willing and able to pay profit-enabling prices. Accordingly, marketing plays a role in diverse activities such as guiding R&D and promoting, pricing, and distributing products.

MARKETING AS A GUIDE FOR R&D

R&D is unpredictable and, while R&D projects may have specific aims, it is not possible to actively control the outcome of R&D efforts. Unlike engineering projects, where a set of blueprints and plans dictate what the final product will be, it is not possible to choose what R&D will discover, and not all R&D products are worth commercializing. The challenges of guiding and predicting the outcomes of R&D create a pressing need to integrate marketing with R&D. Promising leads emerging from R&D must be evaluated for their market potential to ensure proper allocation of resources to promising projects and to avoid wasting time and money on less lucrative opportunities.

The case of Pfizer's blockbuster Viagra is an excellent case for the value of integrating marketing with R&D. In the course of clinical trials for Viagra's original intended market, angina, a side effect of increased erections was discovered. Recognizing a potential to treat erectile dysfunction (a term popularized by Pfizer to replace *impotence* and facilitate marketing the drug), and aware that this market was poorly served, Pfizer proceeded with a small clinical trial with patients suffering from erectile dysfunction and found the drug to be highly effective. Buoyed by an aggressive marketing campaign, Viagra was prescribed more than 4 million times in the first week following FDA approval, and the drug quickly grew beyond $1 billion in revenues.

Beyond guiding R&D, marketing is also necessary to realize the full commercial potential of biotechnology. The notion that novel and useful products and services will attract customers in the absence of marketing is as untrue for biotechnology as it is for any other industry. It is imperative to conduct a thorough market assessment and develop a marketing plan to effectively profit from developing innovative biotechnology products. The market must be characterized in order to establish plans for, and effectively manage, the development and delivery of products and services that satisfy customer needs profitably. The case of Immunex's acquisition resulting from underestimating the market size of its first-in-class drug

---

Box

Biotechnology myth: Build it and they will come

Evacyte Corporation was spun out of the University of Texas at Austin in 2000 to commercialize a patented method for optically stretching cells between two laser beams as an assay for early cancer detection. The physics and science have been published and the technology is capable of correlating changes in the cytoskeleton with metastatic progression. The company won the Austin KPMG Start-up Standout award in 2001 for most innovative start-up. With the company established and technology proven, why did

Evactye go bankrupt?

The original belief was that the optical stretcher technology would be a new diagnostic for cancer and that every pathologist would eventually have one. Evacyte had always foreseen FDA trials as a pre-requisite for market entry. What would turn out to be harder was changing how doctors practice medicine. The premise of Evacyte's technology was that healthy cells don't stretch and cancer cells do, which makes sense if you think about cancer cells multiplying rapidly and needing to break down the cytoskeleton to do so. Pathologists using the optical stretcher, however, would not change how they practiced medicine. If a pathologist has any indication of cancer they treat the cancer as aggressively as possible, so most of the times that a cell did stretch there would be other signs of cancer and therefore the stretchiness didn't help make decisions that insurance would reimburse. If the cells did not stretch they would not withhold some preventative procedure. Without an oncologist prescribing the test there would be no insurance reimbursement, which meant the death of the technology and Evacyte.

Evacyte built a technology that worked, but what was built did not solve a problem in the marketplace. Evacyte had a patented product and a clearly articulated path to FDA approval. Many biotechnology start-ups are likewise based around science that makes an invention or technology possible. What differentiates successful startups is a vision beyond developing science. Successful companies must start with a solid understanding of the market and the pain that is being addressed. It is not enough to achieve FDA approval; successful companies must engage end users and have a compelling case for how a technology will be used before insurance providers or other payers will reimburse the cost.

Many biotechnology entrepreneurs have a "Field of Dreams" attitude to what they are working on and see the path to success leading from the bench top to an invention, device, or drug. What is important in building a successful biotechnology company is to think backwards to how the innovation solves a problem in a defined market that insurance providers are willing to cover and that doctors are willing to prescribe. Without those foundations no matter what you build, *they* will not come.

Contributed by Christian Walker, Founder, Evacyte Corporation. biowalk@hotmail.com

Enbrel is presented in a box later in this chapter.

Market evaluation is also important to fund and justify research efforts. It is the commercial viability of a product that influences the ability to attract funding during development. Furthermore, because revenues derive from the sale of products that have completed expensive and lengthy research programs, it is important to assess marketing issues at the outset of a research project in order to determine, as best as possible, that a profit-enabling market exists.

CRITERIA FOR SUCCESS

It is important to objectively evaluate factors supporting and challenging the profitable sale of biotechnology products and services early in their development. Three important criteria to consider in evaluating market conditions are:

- Freedom to operate
- Availability of technological factors
- Ability to generate a profit

Any criteria that are not satisfied upon initiation of a project will ultimately have to be satisfied in order for the project to succeed. For example, if a market is blocked due to lack of freedom to operate, licensing patents or designing around blocking patents can enable a company to serve a market and may also block competitors. Alternatively, if market access is impeded due to unavailability of technological factors or if existing methods are too expensive to permit profitable sales, performing the requisite R&D can facilitate access to the market, and protecting the R&D with patents or trade secrets can serve as a barrier to competitors.

If all criteria are met at the outset, as is often the case with generic drugs, then the barriers to entry are very low, unless a company has a competitive advantage. This advantage may be in the form of specialized expertise, intellectual property protection, or exclusive access to necessary tools or reagents.

Freedom to Operate

To realize the potential of biotechnology inventions, companies require freedom to operate. Patents play an important role in defining freedom to operate. Competitor's patents may cover processes that limit the market opportunities that a company can target. Additionally, upstream and downstream patents can impede commercialization. For example, if a biotechnology product or service needs to integrate with a patented product or process, requires the use of a patented technique or product, or interferes with third party patents, agreements with the holders of the necessary patents will have to be reached.

Alternatives to licensing requisite patents are engineering around patents, challenging them, or knowingly infringing them. Patent challenge is described in further detail in Chapter 7. In cases where patent challenge or engineering around a patent is impossible or prohibitively expensive, infringement may be the only available option. A significant downside of infringement is that damages may be trebled if a court can be convinced that a patent was deliberately infringed and the infringer is unable to demonstrate invalidity of the patent in question. Government regulations and public support or resistance can also influence freedom to operate; bans on stem cell research funding or transgenic crop planting, for example, have impeded biotechnology company operations.

Technological Factors

Biotechnology product development is a research-intensive endeavor. The bulk of effort is invested in either isolating and producing products, or in developing methods to isolate and produce products. In some cases the raw materials may already exist in a suitable form, whereas in others they must be produced or isolated from natural sources. Additionally, the process to produce a biotechnology product may or may not be known at the outset. Much of the effort in drug development is expended in identifying effective drugs and demonstrating their safety. In many gene therapy applications, for example, the gene that must be delivered for therapeutic effect is known and characterized at the outset. The challenge is to devise a method to deliver it in an effective manner. Conversely, the challenge for monoclonal antibody therapy is to produce and/or isolate appropriate therapeutic antibodies.

PROFITABILITY

Whereas the first two criteria, freedom to operate and availability of technological factors, are necessary to be able to produce a product, the ability to generate a profit is essential for commercial

---

Box

Risks of selling to "ideal" markets

With great opportunity comes many risks. While many would describe an ideal market as large, growing, and underserved, there are some serious pitfalls to selling in such markets. Proactive management is required to assess and manage these risks.

One of the concerns about serving large markets is that unforeseen side effects may emerge. Clinical trials test for safety concerns in large test populations, but patient populations have more variance than clinical trial populations. Furthermore, when a drug is widely prescribed it is difficult to identify off-label use and issue proactive guidance to physicians. This lack of knowledge about the cause of negative responses may lead to a drug being pulled from the market.

Drug interactions and conflicting disease conditions can also cause complications, and are common in growing markets such as treatments for conditions associated with aging, like arthritis, arteriosclerosis, and cancers. For example, the anti-arthritis drug Enbrel interacts poorly with cancer and psychiatric drugs. Additionally, many drugs stress the liver, limiting their use in patients with compromised livers.

There are also reimbursement risks. For example, Genzyme's drugs which focus on orphan diseases—conditions affecting small populations—carry very high prices for several reasons, including the need to cover expensive drug development programs on fewer sales than other drugs. These high prices make Genzyme's drugs a target for cost-cutting measures by health plans.

Adapted from: Friedman, Y. The unexpected risks of the 'ideal' market. In *Strategic Business Risk*. Ernst & Young and Oxford Analytica, 2008.

success. It is important to assess the projected cost of development, time to market, market size, and profit margins to effectively evaluate the merits of a project and select from multiple projects.

The process of satisfying the criteria to produce and profitably sell a biotechnology product or service requires three activities: research, development, and marketing. Research identifies potential lead compounds and technologies; development refines and characterizes these products and technologies in preparation for marketing; and, marketing identifies customer needs and develops methods to reach customers.

REGULATORY AND PUBLIC APPROVAL

A major concern for all biotechnology products is safety. The potential for unwanted effects on the environment and human health motivates government control and public awareness of applications of biotechnology. A unique aspect of the biotechnology industry is that any biotechnology product meant for human use, or which can interact with the environment, must receive regulatory approval.

Regulatory bodies are described in greater detail in Chapter 8. Briefly, the Food and Drug Administration (FDA) is responsible for ensuring that foods, drugs, and their manufacturing processes are safe for human consumption. The FDA also requires drugs to be proven effective and their labeling to be appropriate. The Department of Agriculture (USDA) regulates plant pests, plants, and veterinary biologics to ensure that they will not harm the environment or other plants and animals. The Environmental Protection Agency (EPA) ensures the safety of both chemically and biologically produced pesticides. A benefit of these regulatory burdens is that consumers are provided with objective determinations of the safety and efficacy of biotechnology products.

Beyond government regulations, public approval can also have a significant impact on marketability. Public aversion to genetically modified foods in Europe, partially attributed to a marketing campaign by a British supermarket aimed at displacing imported products, is implicated in the development of formal regulations in many EU nations banning the import of U.S. corn and other genetically

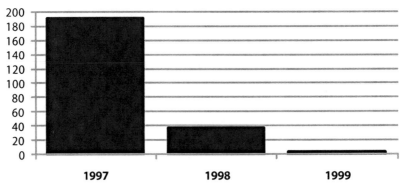

Figure 13.1 *Regulations affect sales: U.S. corn exports following partial EU ban*
Source: USDA Economic Research Service

modified foods. The impact of these regulations, shown in Figure 13.1, has been dramatic.

HORIZONTAL AND VERTICAL PRODUCTS

An important distinction in biotechnology product sales is between products that are sold to companies at a similar position in a production pipeline as the seller and sales to companies at a different position. Drugs, industrial enzymes, and other manufactured products are termed vertical products because they are sold to companies at a different position in the application pipeline, such as distributors and end-users. Horizontal products such as technology platforms, research tools, and reagents are sold to firms with operations similar to the vendor; firms involved in research and development of biotechnology products and services.

Vertical product development is often favored over horizontal product development because of potentially larger markets, higher barriers to entry, and higher gross margins. For vertical product development, the risks of product failure and potentially long development times add substantial risk.

The benefits of horizontal product focus over vertical product development derive from reduced risk and cost of development. Tools generally have fewer regulatory hurdles than vertical products and often face fewer challenges in development. While individual

Table 13.1 *Horizontal and vertical products*

Horizontal	• Tools and reagents • Sold to peers • Low developmental risk, reduced regulatory burden
Vertical	• Drugs, manufactured products • Sold to end-users further down value chain • High developmental risk, high barriers to entry

vertical product companies may succeed or fail, as a group they all require tools to perform their research. Accordingly, horizontal products such as tools and reagents have the potential for relatively stable revenues without the risks associated with vertical product development.

For horizontal products a decreased risk of product development is accompanied by a reduced barrier to entry of competing products. While patent protection and the difficulty of drug development allow drug companies, for example, to sell their products for years without modification, the risk of competition or commoditization requires tool companies to continually refine their offerings. Competitors may develop improved methods that yield better tools or a new technology may emerge, making present offerings obsolete. Additionally, large competitors can devote considerable sums of money to the production of competing tools and engage in prolonged price wars which can overwhelm smaller companies. Strong patent protection of horizontal product technologies is therefore essential to prevent, or at least discourage, competition.

MARKET STRUCTURE AND MARKETING ENVIRONMENT

The purpose of market research is to develop an understanding of the market of a product, the potential customers, and the competitive environment.

Market research typically consists of two components: qualitative marketing research (also called primary market research) and quantitative market research (also called secondary market research). Primary market research involves the collection of original data from sources such as interviews, surveys, product tests, and focus groups. Secondary research is based on existing data from sources such as magazines and newspapers, scientific papers, industry reports, and other publications. Secondary research helps convert raw data from primary research into sales trends, demographic profiles, and business statistics that can help drive strategy.

Ideally, biotechnology companies should seek to serve a market that is large, growing, and underserved. It is important to consider all three criteria in evaluating markets. Large and growing markets permit generous, increasing, revenues, while underserved markets may feature low price elasticity and relatively easy customer acquisition. Asthma therapeutics, for example, represent a large and growing market, but the effectiveness of available medications challenges entrants to differentiate their products and offer a compelling advantage over existing alternatives (for another example, see Box *Inhalable insulin: Holy grail or tiger trap?* in Chapter 12). By contrast, treatments for geriatric ailments have a large and growing underserved market but are challenged by the ability of individuals and health care plans to support profitable sales.

Small companies developing applications for very large markets should consider the implications and possibility of competition from larger and more experienced firms. Small, high quality markets may be preferable to larger ones in certain cases. Competition for popular markets can result in increased price elasticity and limit profits. Furthermore, small companies may find themselves mired in patent litigation—regardless of legal merit—with larger companies.

Products with strong patent protection that are first to market may nonetheless find competition from similar products that bypass patents or from alternative products that serve the same need. Amgen's successful campaign to prevent Transkaryotic Therapies from selling a competing anemia therapy (see Box *Amgen v. Transkaryotic Therapies: Strategic patenting* in Chapter 7), demonstrates how an established company's relatively greater legal expertise and resources can exclude smaller competitors from lucrative markets.

MARKET SEGMENTATION AND TARGETING

Because most biotechnology products are developed for specific markets, targeting is an essential element of any biotechnology product marketing strategy. Targeted marketing is the alignment of marketing efforts for a product or service with its benefits to enhance sales. Market segmentation divides a market into distinct groups of buyers or decision makers. In targeted marketing, each segment of a market is evaluated for its commercial potential and a product's competitive advantages are identified and communicated to each segment. The assumption underlying targeted marketing is that focusing marketing efforts to where a product's unique benefits are valued most enhances commercial success.

The identification and selection of markets, permitting market segmentation and targeting, is facilitated by market research that defines market structures and environments. Once favorable markets are identified, the next objective is to critically evaluate which ones are preferable.

Markets should be measurable, accessible, actionable, and substantial. Measuring the size and value of a market permits an evaluation of the potential for revenue and comparison with other markets. Markets must also be accessible and actionable, because a market that cannot be reached cannot deliver revenues. Finally, a market must be large and profitable enough to warrant targeting. Small biotechnology companies may license shelved drug leads from larger companies because of their ability to profitably serve smaller markets (for an example of drug development for markets without profit potential, see Box *Non-profit drug development* in Chapter 12).

MARKET SIZE

When assessing the size of a market it is not sufficient to simply cite the number of people with a given condition, because it is unlikely that all, or even many, of these individuals will take a given drug. Figure 13.2 shows the progression from determining disease prevalence (the number of people who have a disease) to measuring the size of a target market. For example, while the prevalence of hypertension in the United States is 30 percent—approxi-

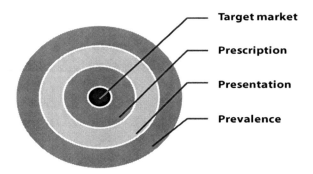

Figure 13.2 *Measuring target market size*

mately 100 million have hypertension—this figure does not represent a target market. Most people are unaware that they have hypertension, and many who are aware do not seek treatment. Even among those who seek treatment, many fail to fill their prescriptions. So, a company producing a hypertension treatment cannot hope to sell their product to 100 million people in the U.S.

To determine the size of the target market, one must consider how many patients are aware that they have a condition, see a physician, obtain a prescription (for the appropriate disease) and fill that prescription. In 2003 there were more than 35 million physician office visits for hypertension, representing a presentation rate of roughly one-third.[1] This figure must still be discounted by the number of repeat visits (how many unique individuals are represented in these 35 million visits), the number of individuals who did not receive prescriptions for hypertension medications, and those who did not fill their prescriptions.

Market sizes may also be a source of contention in licensing, partnership, or funding and acquisition discussions. The potential for market projections to influence the price of an individual product or entire company may lead to disagreements over calculations and methodologies. The Box *Remicade: Resolving valuation disagreements* in Chapter 14 demonstrates how a junior partner can reach agreement on a senior partner's lower estimates without compromising future revenues.

1 Centers for Disease Control

Box

Enbrel: Underestimating market demand

A challenge for first-in-class drugs is that it can be very difficult to assess their potential market size. Immunex's rheumatoid arthritis drug Enbrel was so popular that it set a 24-month growth record for biotechnology drugs. Unable to manufacture sufficient drug to meet demand, Immunex had to resort to drug rationing, initiating an "Enbrel Enrollment Program" to identify and enroll rheumatoid arthritis patients. Immunex's inability to maximize profits during patent-protected sales created an opportunity for competitors to enter the market, reducing the possibility for recovery of missed profits.

While Immunex resolved to retrofit an existing factory to produce more Enbrel, the unfortunate timing illustrates a common challenge faced by drug development firms. In order to meet the initial demand for Enbrel, Immunex would have needed to initiate construction of a manufacturing plant years before FDA approval. The possibility of a late-stage clinical trial failure would have meant that resources necessary to salvage the drug or develop other products might not have been available. The decision of how much manufacturing capacity to build (or lease) therefore requires careful consideration of the likelihood of development success and potential market demand. Immunex's inability to manufacture sufficient Enbrel ultimately cost the company its independence. Deflated stock prices resulting from production shortfalls permitted industry leader Amgen to acquire Immunex.

Strategies to prevent or mitigate production shortfalls include establishing manufacturing partnerships, contracting manufacturing, acquiring manufacturing facilities, or using alternative production techniques. While partnerships and contracting may find unfavorable terms when negotiated under duress, the cost to acquire production facilities may be prohibitive, necessitating acceptance of whatever terms are offered. Alternative production techniques such as transgenic plants and animals are easier to scale, but they face initial development challenges. Because establishing transgenic production methods takes longer and costs more than traditional fermentation methods, many companies may find traditional techniques preferable.

It is also important to understand the factors influencing market growth. Beyond measuring the prevalence (present number of affected individuals) and incidence (annual rate of new cases) of a condition, it is also important to consider penetration into the existing market. Will a drug see rapid adoption, or will sales grow slowly? Are there existing competitors? For what duration will patients take the drug? The Box *Enbrel: Underestimating market demand* illustrates the downside to not being prepared for market growth.

MARKET ENVIRONMENT

In addition to understanding the composition of a market, it is also important to recognize the marketing environment. Market dynamics can exert strong influences on determining the preferred methods to enter and compete in a market. Two commonly used models to evaluate biotechnology markets are PEST analysis and Porter's Five Forces. These frameworks help assess the accessibility and competitive qualities of a given market.

PEST ANALYSIS

The examination of political (including legal and regulatory), economic, social, and technological factors that have an impact on business operations evaluates the market potential and possible hurdles for a product.

Laws governing biotechnology applications, the patent landscape, and the requirements to satisfy regulatory agencies serve to constrain biotechnology applications. The government, through laws and regulatory agencies, can influence which biotechnology applications can be commercialized.

Government-issued patents confer the right to exclude others from practicing an invention and permit pioneers to secure a competitive advantage. These 20-year market exclusivity grants enable companies to secure a return on their research efforts. Other provisions such as the Orphan Drug Act provide incentives for development and grant market exclusivity in the absence of patents.

The regulatory environment for a given application is also an important consideration in defining a market. Products such as drugs and genetically modified foods, for example, face similar regulatory requirements, but the potential profit from drug sales generally exceeds that for agricultural products. This comparison simplifies the multiple factors making drug development such a common theme for biotechnology companies. Conversely, for products with similar revenue potential, the relative challenge of successful development or achieving regulatory clearance may favor some over others.

Regulatory uncertainty is also a significant factor: first-in-class drugs and other innovative products often face uncertain regulatory paths. The potential for time or cost overruns in obtaining regulatory clearance must be considered. Furthermore, bans on specific products such as genetically modified foods, or changes in patent laws, can also have profound effects on the commercial prospects for biotechnology products.

Economic considerations range from the cost of development and availability of funding to the potential for profitable sales. Factors such as lending interest rates and support in public and private markets influence the ability of biotechnology companies to raise funds. While it is possible to bootstrap biotechnology development and sell promising leads or form research alliances to enable commercialization, positive funding environments permit companies to grow faster and achieve greater results. Celera, for example, was able to use public market support to fund its bid to sequence the human genome in advance of an international government-funded effort. Human Genome Sciences leveraged equity offerings to favorable markets in 2000 to expand its infrastructure and capabilities by proactively building the fourth-largest manufacturing plant in the northeastern United States to accommodate the future production of products under development.

It is important to identify the decision makers influencing the purchase of biotechnology products (see *Reimbursement*, later in this Chapter). Physicians and managed care institutions have the greatest influence on purchase decisions for drugs. End-consumers or decision makers earlier in the distribution chain may similarly have selection priority for other biotechnology products (see Figure 13.4). Identification of decision makers and factors in decision making can permit effective targeting of decision makers and appropriate design of marketing strategies.

Social considerations are especially important in biotechnology. Social attitudes toward products and research techniques can have profound implications on the marketability of biotechnology products and the ability of companies to use controversial techniques in their research. Consumer resistance to genetically modified foods led to a ban in Europe. Opposition to animal testing, genetically-modified crop plantings, or stem cell research can also impede R&D. It is also important to consider how products suit the lifestyles of consumers. One of the contributions to the success of Pfizer's erectile dysfunction drug Viagra was its appeal relative to alternative treatments. By contrast, Pfizer's market withdrawal of its inhalable insulin drug was partially attributed to the similarity of the dosage system to drug paraphernalia.

Analysis of technological factors is also important to develop an understanding of biotechnology product markets and guide development efforts. Evaluating the rate of technological change can help assess whether a product is likely to become obsolete shortly after launch or if development should be postponed until technologies to facilitate development emerge. Identification of the technologies that are essential for the use of a product can define the potential market and aid in projecting the conversion costs for customers.

While it is important to consider factors affecting the commercialization of biotechnology products, the ability to develop a product is also an important consideration. The technical challenges of product development have a significant impact on the ability to develop biotechnology products. Whereas engineering projects, for example, may face unanticipated technical challenges that exceed budget estimates, unexpected hurdles in biotechnology projects can result in outright failure. This reality makes a technical assessment vitally important, because of the possibility that R&D expenses may never be recovered.

PORTER'S FIVE FORCES

Porter's Five Forces, developed by Harvard Business School Professor Michael Porter, identify five forces that drive competition within an industry and interact to influence the environment for a market:

- The intensity of rivalry among existing competitors
- The threat of entry by new competitors
- Pressure from substitute products or services
- The bargaining power of suppliers
- The bargaining power of buyers

Enumeration of the relative contributions of each force permits the objective characterization of a market and can also enable the identification and management of factors influencing profitability.

The intensity of rivalry among competing firms is an important contributor to market dynamics. Rivalry is influenced by two counter-balancing elements: barriers to entry and substitute products, and the relative bargaining power of suppliers and customers.

Barriers to entry and differentiation among products influence the potential for competition in a market. Biotechnology products tend to be high-value products, meaning that strong barriers to entry are required to fend off competition. Some of the elements that protect biotechnology markets are patents and other intellectual property protection, high capital requirements, regulatory requirements, and access to sales and distribution channels. Substitutes

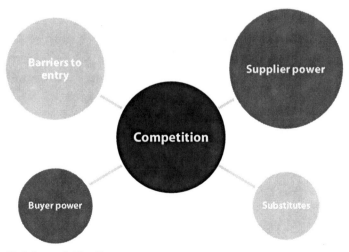

Figure 13.3 *Porter's Five Forces*

such as generic drugs or novel products that do not infringe on intellectual property can circumvent barriers to entry and diminish their effectiveness.

The relative bargaining power of suppliers and customers influences the potential for profitable sales. The power of suppliers is influenced by the number of suppliers in a market and the cost to switch suppliers. The drugs Depakote and Lamictal illustrate how products can reduce buyer power. Both neurological drugs share similar indications, but the combination of both drugs can be lethal. Because switching from one drug to the other requires taking both for a short period of time—a potentially risky prospect—the market of each drug is protected by the difficulty of switching from one drug to the other, reducing the power of customers. Low buyer power can also be a barrier to adoption. One of the challenges in selling platforms such as microarrays is that consumers may be unwilling to make large investments in technologies that bind them to a single supplier, sometimes preferring to develop capabilities in-house at greater cost to ensure future flexibility.

REIMBURSEMENT

Reimbursement is a critical element in commercializing drugs. Patients seldom pay the full price of the drugs which they are prescribed. Third-party payers such as health insurers and government programs usually cover all most of the cost. Healthcare coverage in the United States is provided primarily from three sources: Medicaid (coverage for individuals with low incomes and limited resources), Medicare (individuals 65 years of age and older, some disabled people under 65 years of age, and people with permanent kidney failure), and managed care organizations. The Box *Biotechnology myth: Build it and they will come* earlier in this chapter is a clear example of how failure to consider reimbursement can prevent commercialization of an approved product. FDA approval alone is not sufficient to ensure sales; payers must also be willing to reimburse patients for a drug.

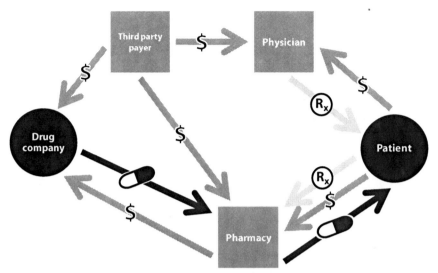

Figure 13.4 *Simplified drug market model*

CONSUMERS, GATEKEEPERS, AND PAYERS

The ethical drug market is a complex ensemble of consumers, gatekeepers, and payers. The parties responsible for selecting drugs, using them, and paying for them each have different motivations and needs. The challenge for firms serving this market is to identify the various types of customers and meet their specific needs.

Figure 13.4 shows a simplified model of the flows of money and drugs in the ethical drug market. In this model, drugs are supplied to the pharmacy from the manufacturer, often through a series of wholesalers and distributors. Patients are given prescriptions by a physician and take the prescriptions to a pharmacy where a pharmacist dispenses the requested drug. Underlying this series of drug and prescription transactions is a network of entities and incentive mechanisms that ultimately fills prescriptions and compensates all the players. Physicians are compensated by third-party payers (insurers, Medicare, Medicaid, etc.) and co-payments from patients. Pharmacies are also compensated by third-party payers and co-payments from patients.

Third-party payers exert influence over prescription patterns by dictating which drugs are reimbursed, which drugs require prior authorization for prescription, which substitutes should be applied,

Box

Avastin: Controlling off-label use

Lucentis and Avastin are two monoclonal antibodies produced by Genentech. Lucentis is FDA-approved for advanced, age-related macular degeneration (AMD) and Avastin is approved for colon and other cancers, but not for macular degeneration. Both drugs operate by similar mechanisms and Avastin substitutes favorably for Lucentis, which motivated many physicians to use Avastin in place of the far more expensive Lucentis. Citing safety concerns, Genentech took steps to control off-label prescription of Avastin for AMD.

To control off-label prescription of Avastin, Genentech announced that it would no longer permit direct purchases by compounding pharmacies—pharmacies that repackage the drug into smaller units for eye injection. Genentech cited FDA concerns regarding the safety and efficacy of drugs from compounding pharmacies, but the price difference between the two drugs—Avastin costs less than $150 per treatment and Lucentis costs more than $2,000 per treatment—led many to question Genentech's motivations. In response to protests from physicians that Genentech's actions could deprive many patients from accessing the vision-saving drug, the National Eye Institute of the National Institutes of Health announced a multi-center clinical trial to test the effectiveness of Avastin in treatment of AMD.

and how much co-payment is required. A drug's pharmacoeconomic profile, or cost-benefit ratio, can have a profound effect on prescription patterns. Third party payers are motivated to reduce their healthcare expenditures. Drugs that can shorten hospital stays, substitute for expensive treatments, prevent diseases which are expensive to treat, or otherwise reduce the cost of health plan administration are therefore able to command premium prices.

Physicians are the ultimate gatekeepers, deciding which drugs to prescribe. The physician's primary consideration is the expected clinical improvement of the patient. In addition to safety and efficacy, physicians must also consider factors such as administration and monitoring. Drugs which place a burden on healthcare professionals may be reimbursed at the same rate as simpler alternatives, creating

a financial disincentive to their prescription. Side effects and toler-
ability place a burden on a physician's time. The need to monitor
side effects and clinical progress through repeated visits or laboratory testing further increase this burden. Accordingly, a drug with
a high maintenance regimen or even a low incidence of particularly
dangerous side effects may be less preferable to a lower-maintenance
drug with a more predictable side effect profile (see the examples of
Herceptin and Aczone in Box *Personalized medicine and drug sales*
in Chapter 6).

Patients also have an impact on drug sales. To visit a physician
and obtain a prescription, patients either need to suffer an incident
which brings them to a hospital, or otherwise realize they have a
condition for which a treatment may exist and feel compelled to seek
treatment. Patient education programs are commonly used to encourage potential patients to consult physicians regarding treatment
options.

Patients also have a role in the process of obtaining prescriptions and choosing to fill them. If the cost of a filling a prescription
is too high (e.g. due to high copayments or lack of reimbursement),
the quality-of-life impact insufficiently motivating, or the possible
side effects too daunting, a patient may simply elect to not fill a prescription. Furthermore, patients may exert influence over their physician's prescription through a preference for delivery methods. For
example oral administration and metered patches are generally preferred over injections.

DRUG COVERAGE DECISIONS

Drug coverage decisions are based on a combination of clinical and economic considerations. Payers can use several methods to
discourage or favor prescription of individual drugs. They generally
seek evidence that a drug will improve patient outcomes or reduce
costs. Formulations that improve patient experiences, for example,
may not be favorably covered without a compelling clinical or cost-
saving result.

A recent study of coverage policies for several high-cost drugs
examined the coverage decision-making processes of 53 managed

care organizations.[2] The most common strategies to limit medication use are to limit the quantity or duration of medication use, or to exclude medical conditions or drug classes. The next most common strategy to limit use was requiring prior authorization. Data sources used to support coverage decisions included literature reviews by internal experts, information from pharmacy benefit managers (e.g., Medco and Express Scripts), and external experts such as the Blue Cross/Blue Shield Technology Evaluation Center.

The Blue Cross/Blue Shield Technology Evaluation Center uses the following criteria to assess whether a technology provides sufficient health benefits:

1. The technology must have final approval from the appropriate governmental regulatory bodies
2. The scientific evidence must permit conclusions concerning the effect of the technology on health outcomes
3. The technology must improve the net health outcome
4. The technology must be as beneficial as any established alternatives
5. The improvement must be attainable outside the investigational settings[3]

TEC assessments are used by public and private sector clients to make coverage decisions. Cost is reportedly not the sole consideration in coverage decisions. FDA approval, and availability of alternatives were cited as additional factors.[4] Requests for coverage are also standardized. A majority of payers use the Academy of Managed Care Pharmacy's Format for Formulary Submissions, a standardized methodology for assessing drugs based on the value they provide.[5]

---

2 Titlow, K., Randel, L., Clancy, C.M., Emanuel, E.J. Drug coverage decisions: The role of dollars and values. *Health Affairs*, 2000. 19(2):240-247.

3 Blue Cross and Blue Shield Association. http://www.bcbs.com/betterknowledge/tec/tec-criteria.html

4 Titlow, K., Randel, L., Clancy, C.M., Emanuel, E.J. Drug coverage decisions: The role of dollars and values. *Health Affairs*, 2000. 19(2):240-247.

5 Academy of Managed Care Pharmacy. http://www.amcp.org

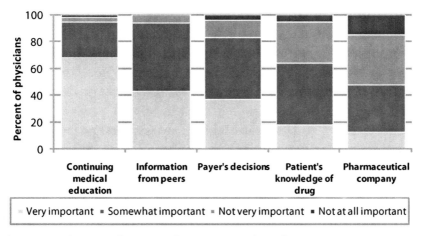

Figure 13.5 *Factors influencing physician prescribing decisions*
Source: Outlook 2008. *Tufts Center for Drug Development*, 2008.

CONTROLLING PLAN COSTS

Cost-controlling measures must be balanced against the mission of healthcare payers; they ultimately are in the business of facilitating access to healthcare. Beyond excluding entire disease categories or drug classes and requiring prior authorization, managed care organizations may also attenuate prescription preferences through co-payment tiers. The three basic tiers are Tier 1, no co-payment; Tier 2, low co-payment; and, Tier 3, high co-payment. A fourth tier, Tier 4, requires patients to pay a percentage of drug costs. Payers explain that these tiers give patients subsidized access to drugs which might otherwise not be covered, and also enable patients to chose alternatives with less favorable dosage frequencies and methods, or reduced efficacy, in order to reduce personal expenses.

MARKETING PLANNING AND STRATEGY

ROUTE TO MARKET

Once markets have been defined and evaluated, the next objective is to position offerings and develop a route to market.

Designing a successful marketing strategy requires a thorough characterization of a product's comparative and competitive advantages and disadvantages. This insight should derive from product-specific primary market research. Developing data and insights from primary research requires an understanding of the marketing environment. Important factors in evaluating a marketing environment include identifying the decision makers, defining the decision process, determining how a product matches decision criteria, and assessing how a product can fit or change practice patterns.

Product and market characteristics are important considerations in developing sales strategies. An important factor enabling communication with consumers is the nature of sales and distribution channels. Partnering with a firm with an established sales force can reach a large number of customers but faces the possibility that the sales partner may neglect an external product for more profitable internal products. In-house sales forces benefit from specialized training and a focus on company products. Whereas external sales forces may have greater reach and established credibility, they may also lack motivation or detailed knowledge of offerings. Additionally, while a large customer base may be best served by a large sales force, complex products may necessitate a small, well trained sales force.

The selection of distribution options is also important. Distributors can facilitate delivery of products to consumers, but the need to compensate distributors may increase retail prices and reduce the share of revenue and market intelligence available to the producer. A benefit of direct distribution is the ability to interact directly with customers or terminal suppliers and gain first-hand knowledge of product perceptions and purchase patterns.

GENERATING SALES BEFORE MARKETING AUTHORIZATION

There are two ways in which drug developers can generate sales before FDA approval and marketing authorization are granted, both leveraging differences in drug approval between different countries.

For a given drug, approval may be faster or slower in various countries (for further discussion, see *Regulation* in Chapter 16). If a drug can gain approval in one market prior to approval in a second, it is possible to generate significant revenues from these initial sales. Having a patient population paying to receive a drug can also have benefits for second market approval as it provides a reduced cost method to monitor safety and efficacy, and can help in the design of clinical trials.

Even if a drug is not approved in another country, it may be possible to generate revenues from compassionate use programs. The U.S. compassionate use programs, described in Chapter 8, enable limited distribution of drugs prior to marketing approval. A differentiator of European Named Patient Programs, which are similar to U.S. compassionate use programs, is that unlicensed drugs can be reimbursed under Named Patient Programs, providing an opportunity to generate revenues from drugs still under development. Named Patient Programs can also speed post-launch uptake, as they increase physician familiarity with drugs prior to launch.[6]

LIFECYCLE MANAGEMENT AND MARKETING MIX

Traditional product lifecycles are often drawn as bell-shaped curves (see Figure 13.6, top panel). For innovative products such as patented drugs or those facing obsolescence when improved alternatives emerge (such as bioinformatics software and databases, or research tools), product adoption rates can be relatively faster and decline rates more precipitous (see Figure 13.6, bottom panel), although the absence of clear guidance on generic biologic approval can prevent price competition following patent expiration.

As a product goes through the life cycle of development; introduction; growth; maturity; and, decline and/or patent expiration, there are different strategies that influence marketing activities for the product. For example, for discovery through growth, intense marketing is required to get the most out of the investment in the product. As a product matures and declines, less is spent and higher returns are sought on the existing spend. Usually when a product goes off patent, marketing spend and sales support are withdrawn as alternatives (e.g. generic drugs) enter the market and capture market share based on discounted price strategies. Strategies to delay or offset market decline include patent life extensions (described in Chapter 7) and R&D strategies to develop brand extensions (described in

6 Emmer, Gene. President, Med Services Europe. Personal interview. October 2005.

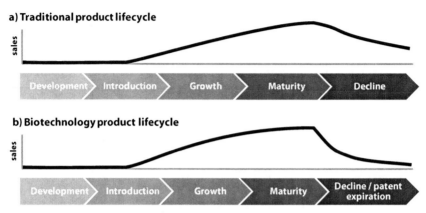

Figure 13.6 *Biotechnology product lifecycle*

Chapter 12).

The distinct lifecycle stages also influence the set of marketing tools and tactics, also known as the marketing mix, used to optimize the distribution of the marketing budget to maximize sales. Whereas early-stage activities may focus on product positioning and reinforcement of value messages, later-stage activities may favor brand awareness to preempt generic and encourage a shift to follow-on products.

Chapter 14
Licensing, Alliances, and Mergers

It sounded like an excellent plan, no doubt, and very neatly arranged; the only difficulty was, that she had no idea how to set about it.
Lewis Carroll, Alice's Adventures in Wonderland

Pharmaceutical firms historically fully integrated all activities from drug discovery through development and manufacturing, and early biotechnology companies sought to grow to the same breadth. Numerous market forces have led pharmaceutical companies to outsource many activities, focusing on areas like clinical trial management and marketing, while biotechnology companies have likewise specialized in discrete elements of product discovery, development, and commercialization.

To maintain focus on core activities and benefit from the wealth of specialists who can fill development needs, companies can use licensing, outsourcing, and alliances. Following successful research and development, young biotechnology companies should maximize their revenue by manufacturing and selling products and services themselves, licensing them to others, or combining both strategies by dividing markets geographically or by therapeutic segments. One reason why smaller firms may choose to ally with established firms is for their extensive and experienced sales and marketing forces that may be essential for success. These relationships can enable a company to perform activities for which it lacks expertise or which are better performed by partners possessing economies of scale.

Mergers, acquisitions, and spin-offs are alternative means of obtaining expertise and capabilities, and maintaining focus. Estab-

Table 14.1 *Inter-company transactions*

Licensing	Acquiring the rights to a technology.
Outsourcing	Contracting defined tasks to outside parties.
Alliance	Partnership to share in the development and commercialization of a product. Usually involves licensing of technologies.
Merger / Acquisition	Formal combination of two companies.
Spin-off	Formation of a new firm from a larger one.

lished companies may elect to acquire smaller companies rather than engaging in licensing or alliances, and small companies may elect to merge to build bulk. Companies may likewise decide to spin-off commercial opportunities tangential to the core mission, enabling the spin-off company to obtain independent financing and shield the parent from high R&D costs and risks which shareholders may not wish to bear.

There is some variance in the definition of terms used to describe some of these inter-company transactions. The term "license" refers to a an agreement whereby one party gains access to another's technology (e.g., a patent license). Licenses are also involved in more complex collaborations and developmental partnerships, but for the purposes of this chapter, licensing describes a simple cash or royalty agreement in which one company gains access to a defined, existing, technology from another. Alliances are similar to licensing deals, with the addition of shared risk and responsibility over the development of a product. Unlike licenses, which facilitate the simple transfer of existing technologies, alliances enable participation in the development of a technology, which the developer generally licences to the sponsor upon completion. License fees and payments for achieving development milestones may be spread over the development of a product, enabling the sponsor to reward the developer for progress, while limiting their own risk exposure.

LICENSING AND OUTSOURCING

A biotechnology company's intellectual property can generate revenue indirectly through product sales or directly through technology licenses. An important consideration in licensing is identifying and negotiating areas where buyers and sellers have conflicting goals, such as cost issues, strategic concerns of the licensor and licensee, competition risks, and exclusivity issues.

Granting licenses gives companies an opportunity to test and develop applications while also validating the value of patents. Several genomics and molecular evolution companies have used this strategy to demonstrate the value of their technology platforms while retaining specific opportunities for internal development.

Licensing technology platforms for non-competing interests also permits licensors to gain prestige from the achievements of licensees, while generating revenues from the licenses. For example, in the mid-1990s Dyax held lucrative patents that covered a widely-used technology, yet found itself in a poor financial position. Without sufficient resources to pursue patent infringement suits, and recognizing that the widespread use of their technology granted the company esteem, it was decided that rather than prosecute infringers, Dyax should extend undemanding licensing terms. By leveraging the popularity of their technology and assembling a long list of top-tier companies on its list of licensees, Dyax was able to gain a great deal of credibility with investors.

LICENSE TYPES

A survey of the license types utilized by the National Institutes of Health Office of Technology Transfer (NIH OTT) provides an overview of licensing options.[1] An important distinction between the NIH and commercial licensors is that the NIH OTT's principal mission is the timely introduction of new products and technologies into the marketplace, to ensure that NIH R&D can serve the pubic good. Most commercial enterprises, being funded by revenues rather than taxes, place an emphasis on revenue generation.

---

1 Feindt H.H. 2007. Administration of technology licenses. In *Intellectual Property Management in Health and Agricultural Innovation: A Handbook of Best Practices* (eds. A. Krattiger, R.T. Mahoney, L. Nelsen, *et al.*). MIHR: Oxford, U.K., and PIPRA: Davis, U.S.A.

Commercial evaluation licenses

The purpose of these evaluation licenses, also known as options, is to enable companies to evaluate a technology for a limited amount of time without the burden of deciding on, and meeting, terms for a long-term patent commercialization or internal-use license. These evaluations can streamline licensing negotiations by reducing risk and allowing a licensor to define the utility and scope of a technology.

Patent commercialization licenses

These licenses may be exclusive (granted to only one party) or non-exclusive (granted to more than one party). The benefit of an exclusive license, as described below, is that the licensee can control a market and prevent others from practicing the invention. Exclusive licensees must often share the burden of pursuing patent infringers, and may also face license withdrawal should they fail to sufficiently advance research or commercialize the technology. Nonexclusive licenses may be preferred if a patent has broad uses—beyond the ability of a single party to fully exploit them all—or if multiple parties can utilize licenses non-competitively.

Nonexclusive patent licenses for internal use

Licenses for internal use are best suited for platform technologies which may enable commercialization, but which are not marketable components. In such cases nonexclusive licensing enables multiple parties to develop products, enhancing the overall commercial impact of the technologies (see Box *Cohen-Boyer: Broad licensing to maximize revenues,* below).

Biological materials licenses

Licenses for biological materials provide access to nonpatented materials which may be very expensive to prepare or only be available from the laboratory that made them. The purpose of these licenses, like commercialization licenses, is to facilitate the translation of NIH technologies to impact the public—either by providing licensees with valuable reagents, or by licensing a commercial partner to directly sell useful biological materials produced by NIH scientists.

Table 14.2 *Typical NIH license obligations*

	Evaluation	Exclusive patent for commercial use	Nonexclusive patent for commercial use	Nonexclusive patent for internal use	Biological materials for commercial sale	Biological materials for internal use
			License type			
License execution fees	■	■	■	■	■	■
Annual royalties	▨	■	■		■	
Past patent-prosecution fees		■	▨			
Ongoing patent-prosecution and patent-maintenance fees		■	▨			
Reports on commercial development or research progress	■	■	■		▨	
Report of performance benchmark achievement		■			■	
Performance benchmark royalties		■				
Report of first commercial sale		■			■	
Earned royalties on product sales		■	■		■	
Sublicensing royalties		■				
License renewal or term extension fees			▨	■	■	■

Key:

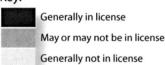

■ Generally in license

▨ May or may not be in license

□ Generally not in license

Source: Feindt H.H. 2007. Administration of technology licenses. In *Intellectual Property Management in Health and Agricultural Innovation: A Handbook of Best Practices* (eds. A. Krattiger, R.T. Mahoney, L. Nelsen, et al.). MIHR: Oxford, U.K., and PIPRA: Davis, U.S.A.

Software licenses

Software licenses are similar to biological material licenses with the distinction that they provide access to software produced in NIH laboratories. The purpose, as with biological licenses, is to leverage the work of NIH scientists to serve the public.

Box

MedImmune v. Genentech: Licensees gain power

Patent licenses can be viewed as an agreement in which the licensor agrees not to sue the licensee for patent infringement, and the licensee agrees to pay compensation for not being sued. Prior to *MedImmune v. Genentech*, licenses also provided licensors an assurance that licensees would not challenge their patents. To challenge a patent the challenger must be accused of infringement, or in fear of being accused of infringement. Licensees wanting to challenge a patent would have needed to violate the licensing agreement, infringing the patent, in order to challenge a patent. If the licensee lost their challenge they risked paying trebled damages for willful infringement.

The case of *MedImmune v. Genentech* saw the ability of the licensors to prevent licensees from challenging their patents diminished, increasing the bargaining power of licensees. In 1997 MedImmune licensed a Genentech patent for several of its antibody products. The relevant Genentech patent was not issued by the PTO until 2001, at which point Genentech requested additional royalties for one of MedImmune's new products, a near-blockbuster called Synagis. Perceiving Genentech's request as a threat to sue for infringement, MedImmune countered by challenging the validity of Genentech's patent and claiming that it was not infringed by Synagis. Not wanting to face trebled damages, MedImmune paid the requested royalties under protest while it challenged Genentech's patent.

Genentech argued that MedImmune's simultaneous fulfillment of license terms and patent challenge was a breach of the *quid pro quo* that gives licensors protection against challenge in exchange for agreeing not to challenge licensees for infringement. In January 2007 the Supreme Court ruled that MedImmune had the right to challenge Genentech's patent while paying royalties under protest, enabling licensees to challenge patents while simultaneously meeting license terms—avoiding the prospect of trebled damages for willful infringement. The underlying patent dispute was resolved in June 2008 without disclosure of details.

This ruling gives licensees more power in licensing negotiations, because licenses no longer protect licensors against patent challenges. Accordingly, licensors have a greater motivation to offer more favorable licensing terms to discourage licensees from pursuing patent challenges.

Exclusivity

Licenses may be exclusive, granting rights to only one party, or non-exclusive, granting rights to multiple parties. The decision to offer an exclusive or non-exclusive license must consider the benefits of protection from competition for the licensee, and the licensor's desire to maximize license fees and royalties.

Licensees often request exclusive licenses, granting them rights to a technology and barring others. Because of the smaller market—an exclusive license can only be offered to a single party—exclusive licenses tend to be more expensive than non-exclusive licenses, which may motivate cost-conscious licensees to prefer non-exclusive licenses. Additionally, if exclusivity does not confer a significant competitive advantage, as is often the case with technology platforms,

---

Box
Cohen-Boyer: Broad licensing to maximize revenues

A classic example of non-exclusive licensing involves the Cohen-Boyer patents to gene splicing technology, the invention that spawned the biotechnology industry (see Section *Application* in Chapter 2). The motivations to issue the non-exclusive licenses were manifold: the invention predated the Bayh-Dole Act, which granted universities the right to issue exclusive licenses; the broad utility of the patented technology favored license to as many parties as possible; and, expectations within the scientific community and society also favored broad access—non-profit organizations were never required to obtain a license.

The initial license terms, which were modified over the life of the patents, were a $10,000 up-front fee with a $10,000 annual advance against royalties. 468 companies licensed the Cohen-Boyer patents and ultimately generated more than $35 billion in sales of an estimated 2,442 novel products, delivering over $250 million in licensing revenues to Stanford and the University of California.[1]

---

1 Feldman, M.P., Colaianni, A., Liu, C.K. Lessons from the commercialization of the Cohen-Boyer patents: The Stanford University licensing program. In *Intellectual Property Management in Health and Agricultural Innovation: A Handbook of Best Practices* (eds. A Krattiger, RT Mahoney, L Nelsen, et al.). MIHR: Oxford, U.K., and PIPRA: Davis, U.S.A.

there is little reason to request it.

Licensors must also balance the relative benefits of the two license types. Non-exclusive licenses may enable revenues from multiple parties, but in some cases may also diminish the value of the license. Some licenses, such as the rights to a specific drug, are most valuable when only one party has access to them. Exclusive licenses also align the interests of licensor and licensee in the pursuit of infringing parties. A resource-poor licensor may be able to enlist the support of an exclusive licensee in challenging infringers, whereas licensees of a widely-licensed patent have little incentive to support the licensor. A potential downside of exclusive licenses is that a royalty-paying licensee may fail to fully exploit the license and, in the absence of reversion or minimal-performance clauses, the licensor will be unable to fully realize its profit potential (see Box *Exubera: When your partner doesn't sell* later in this chapter).

A solution to resolve the conflict between exclusive and non-exclusive licensing is to offer exclusive licenses for specific markets, defined by geography or application. This compromise gives licensees exclusivity in individual markets and can maximize utilization by aligning licenses with the market strengths of licensees. A common implementation of this strategy is for a small biotechnology firm to retain rights to their domestic market and license all other global rights to a multinational firm. This enables the licensor to focus on the market they are best able to serve while receiving royalties from sales in markets they are unable to reach.

PAYMENT TERMS

The most common payment terms for licenses are fixed payments and royalties. License agreements may also use a combination of the two.

Fixed payments may be assigned in several ways. The simplest is a one-time fee. This may be the preferred option if a license is only needed for a short duration (e.g., a drug screening technology), or if a licensor is strapped for cash and is willing to sacrifice future revenues for a greater one-time sum. Other options include annual fees or, in the case of alliances (described below), milestone payments for developmental progress.

Royalties are calculated based on factors such as net sales or profits, and enable the licensor to share in the profits from commercialization of the licensed technology. A downside of royalties, as described above, is that the licensor's revenues are dependent upon the performance of the licensee. A significant upside to royalties is that the income from royalties is often without cost. If development costs have already been incurred and the royalty-receiving firm does not have to do any further work to generate royalty income, the burden of manufacturing, marketing, and sales, which are recurring costs, fall on the royalty-payer. Accordingly, while gross royalty payments may be lower than milestone payments, when the costs to earn these payments are considered, the net proceeds from royalties may exceed those of milestone payments.

OUTSOURCING

Outsourcing is another means for companies to gain access to technologies. Rather than licensing the rights to use a technology, a company may simply purchase a service based on that technology. Outsourcing also describes other transaction-based collaborations which do not necessarily involve proprietary technologies, such as manufacturing, distribution, sales, etc. The Box *Licensing technology or keeping it in-house* describes some of the considerations in selecting a licensing or outsourcing strategy.

Outsourcing, like alliance formation, is motivated by the need to focus on core competencies, access specialized tools and expertise, and reduce costs. Outsourcing to a firm with the necessary resources, expertise, and economies of scale can help surmount barriers encountered during development. The extreme example of outsourcing is seen in virtual companies—firms that outsource all or most of the elements of research, development, and marketing (virtual companies are described in further detail in Chapter 15). Strategic outsourcing can also off-load non-core activities, enabling companies to focus on core activities.

Primary candidates for outsourcing are operations with short-term discrete deliverables, or those where the duration, pace of technological change, need for specialized expertise, or infrastructure

Box
Licensing technology or keeping it in-house

The choice between licensing a technology platform and permitting licensees to develop applications internally, or keeping a technology in-house and developing and manufacturing products for others requires consideration of implementation practices, intellectual property control, and long-term objectives.

The differing strategies employed by Applied Molecular Evolution (now owned by Eli Lilly) and Diversa (now Verenium) illustrate these differing arrangements. Both firms had patented techniques for molecular evolution: modifying proteins to refine their functions. They differed in their choice to license their technology or keep it in-house.

Applied Molecular Evolution licensed its technology to partners, enabling them to apply the proprietary methods to independently generate variants of defined proteins in their own labs. Diversa took a different tack and did not license their technology, opting instead to generate protein variants in-house for its partners.

The decision to not license technologies may be necessary when significant technical expertise is required to practice a technique. Additionally, keeping expertise in-house creates an obstacle for those wishing to use a technique without license. Conversely, doing work for partners may discourage control-conscious customers and carries the additional burden of distracting company operations from long-term goals.

and recurring expenses favor implementation by outside parties. Drug screening and toxicity testing, for example, where the outputs are often more valuable to a firm than having expertise in-house, may be best performed by specialized firms that can maintain cutting-edge equipment. Activities such as manufacturing and clinical trial management, where specialized skills are required and where set-up times and infrastructure and operating costs can be significant, may likewise be best performed by specialized contractors focusing on those elements and possessing economies of scale. Operations requiring extensive refinement or where the outputs cannot be discretely defined at the outset may be better performed internally, providing greater control over costs and results.

For all the flexibility gained from outsourcing, there are several potential drawbacks. As a greater number of functions are outsourced, a company becomes more reliant upon an increasing number of external factors, and the task of choosing partners and defining and ensuring delivery of agreed service levels becomes more complicated. Responsibility for product quality is also a very important issue. Although contractors may perform actual research, development, and manufacturing, the responsibility for product quality ultimately lies with the outsourcing firm (see the example of Baxter's issues with contaminated heparin in the section *Manufacturing* in Chapter 16). Furthermore, sharing of proprietary technology and trade secrets with a contractor may result in the loss of intellectual property. Methods to protect intellectual property are to outsource only selected manufacturing steps, or to outsource only mature products.

ALLIANCE PARTNERSHIPS

A number of factors make it preferable for biotechnology companies to specialize in discrete elements of the product development pathway. Rather than developing all the abilities necessary for product development and commercialization internally, companies may form alliances with well-positioned partners. Alliances enable companies to acquire the complementary services, technologies, and products necessary for commercial success.

Constant scientific innovation means that biotechnology companies must continually adapt. Unfortunately, progress is challenged by the inability to predict which promising projects will succeed. Alliances enable companies to minimize commitment of resources to individual projects by spreading risk among multiple projects. Virtual integration is possible through alliances, saving infrastructure costs and allowing elements of research, development, and commercialization to be performed by specialists. These strategic alliances can speed time to market, reduce cost of development, and enable the realization of products that might not be possible without contribution from multiple sources. Table 14.3 shows some representative

Table 14.3 *Representative biotechnology alliances*

Partners	Project
Eyetech / Pfizer (2004)	Co-development and co-promotion of Macugen
deCODE / Roche (2001)	DNA probes for novel disease markers
Millennium / Bayer (1998)	Drug targets
Human Genome Sciences / SmithKline (1993)	Rights to products resulting from HGS' planned genome sequence database

Source: Recombinant Capital

biotechnology alliances. Note the progression from simple transfer of drug leads in the 1990s to co-development and co-promotion in the 2000s.

The distinction between alliances and licensing or outsourcing is in the sharing of goals and objectives. In alliances, partners share common goals and objectives. Shared risk and reward systems permit serving partners to benefit from the success of receiving partners. While the distinction between alliances and licensing or outsourcing may be difficult to discern in some cases, licensing and outsourcing are generally contractual relationships where transactions are based on defined products or completion of defined tasks rather than being related to the ultimate success of a project in development.

MAINTAINING FOCUS

As described in the section *Project Selection* in Chapter 12, biotechnology companies must maintain focus to succeed. Having too many diverse aims, or having poorly-defined aims, seldom yields success and can leave a company without sufficient resources to capitalize on promising opportunities.

Because biotechnology companies form around proprietary technologies, selling specialized services to partners can build a stable revenue base which can fund internal research projects. Start-up companies often develop technologies with a broad range of potential uses that extend beyond their research focus. The need to focus efforts on a defined application in order to obtain funding, manage resources, and develop a marketing strategy means that some of these applications will not be pursued. If a proprietary technology can be used for applications that do not compete with a company's goals, that technology can be offered to partners or licensed to other firms, enabling the innovator to partially realize the value of an application that might otherwise be shelved.

Alliances permit partners to focus on their relative strengths and core competencies to achieve their goals more quickly and inexpensively than otherwise possible. While alliances may form between similarly large or small partners with motivations to cooperate on a project, alliances between differently-sized partners are more common. These strategic alliances aim to complement each partner's size-related strengths, such as the ability to innovate or develop and market products.

By partnering with smaller companies, established companies can leverage their marketing, distribution, and sales abilities to profit from the innovation strengths of emergent firms. Such partnerships also provide growing companies access to the financial resources, manufacturing, and distribution networks of mature companies. Tapping the resources of an established partner may also enable products developed by an emergent firm to clear otherwise insurmountable barriers, facilitating commercialization and the realization of markets that would not otherwise be accessible.

The size of large companies, and the constant scrutiny placed on the performance of public companies, also encourages them to form alliances rather than perform R&D in-house. Large public companies can see their stocks fall sharply on bad news, and executives may lose their jobs over high-profile failures. For these firms, formation of alliances with smaller firms which are better-able to absorb risk may be preferable—even if the financial cost is greater—than performing R&D in-house.

Figure 14.1 shows the contribution of alliances to drug development and the rate of late-stage failures in drugs developed by biotechnology companies, pharmaceutical companies, and alliances. On initial examination it may appear that biotechnology companies are far less effective than pharmaceutical companies at developing drugs, but the high rate of late-stage development failure among biotechnology companies is largely attributed to under-funding, not poor development. Another factor contributing to the high propor-

Drugs Approved

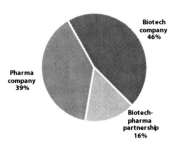

Phase III Failures

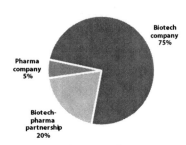

Figure 14.1 *The contribution of alliances to drug development from 2006-2007*
Source: Czerepak, E.A., Ryser, S. Drug approvals and failures: implications for alliances.
Nature Reviews Drug Discovery, 2008. 7:197-198.

tion of late-stage failures among biotechnology companies is that the majority of drugs produced by biotechnology companies in the time period examined were novel products, whereas those produced by pharmaceutical companies were modifications of approved drugs.[2] The rate of failure is naturally expected to be far higher for novel drugs than for modifications of approved drugs.

Funding Growth

Alliances can be a key source of funding for biotechnology companies. Ligand CEO David Robinson estimates that a biotechnology company obtains funding in a 10/50/40 ratio in its first decade: 10 percent of funding comes from venture capital, 50 percent from alliances, and 40 percent from public markets.[3]

Commercialization of recombinant human insulin, the first biotechnology product, is a case example of effective use of alliances. Just two years after starting up, Genentech entered into a licensing and marketing agreement with diabetes market leader Eli Lilly, sacrificing future revenues to support development of recombinant human insulin. Cash reserves from licensing, milestone, and royalty payments later permitted Genentech to develop sufficient internal resources to market its own products. Today Genentech is one of the largest biotechnology firms. It is interesting to note that Genentech

2 Czerepak, E.A., Ryser, S. Drug approvals and failures: implications for alliances. *Nature Reviews Drug Discovery*, 2008. 7:197-198.

3 Hess, J., Evangelista, E. Pharma-biotech alliances. *Contract Pharma*, September 2003.

Table 14.4 *Alliance payment types*

Upfront cash / equity	Payment not requiring specific deliverables from the licensee.
Loan	Cash payment that must be repaid. May be convertible into equity.
R&D funding	Funding based on research and development progress. May not require specific milestones.
Milestone	Funding attached to specific deliverables such as successful clinical trials.
Royalty	A percentage of product revenues.
Quid (*quid pro quo*)	Exchange of benefits 'in-kind' rather than cash or equity.

built its bulk on alliances, not acquisitions—Genentech's first acquisition occurred more than 30 years after formation.

Another example of the use of alliances to help a company grow is Amgen's development of Epogen, the first biotechnology blockbuster drug. To fund Epogen development Amgen partnered with Ortho Biotech, a Johnson & Johnson subsidiary. Amgen received $1 million in an upfront payment in 1985, $6 million in milestone payments over the course of development, and 10 percent royalties for rights to all indications except for kidney dialysis and diagnostics in the United States, and rights for all indications in all other countries except Japan and China (Amgen had an existing partnership with Japanese Kirin Brewing Company). By forgoing rights to global sales—a market Amgen was poorly positioned to serve—Amgen was able to fund development and develop Epogen into a billion-dollar drug. Johnson & Johnson was likewise able to leverage its global sales force to sell Eprex (Amgen sells epoetin alfa as Epogen, Ortho sells it as Procrit in the United States, and Johnson & Johnson sells it as Eprex internationally) globally, with Amgen receiving 10 percent royalties.

ACCESSING RESOURCES AND EXPERTISE

Another motivation for a nascent firm to seek a partner is to support regulatory approval. The time and cost of pre-clinical development, clinical trials, and commercialization may be prohibitive for resource-poor start-ups. Selecting a partner with proven regulatory and clinical expertise can facilitate development and ease regulatory approval.

The challenge of developing techniques, expertise, and resources for product manufacturing may also outstrip a company's resources. Limited resources, the need to remain focused, or the requirement to quickly ramp-up production may also motivate a company to seek manufacturing partners.

Box
Remicade: Resolving valuation disagreements

A contentious issue in negotiating a licensing, alliance, or merger agreement is the value of assets in development. In deals involving drug development, for example, the two parties may disagree on the value of drugs which have yet to receive regulatory approval. How much will development cost? What is the market size? How strong is the intellectual property?

As described in Chapter 11, valuation methods are subjective, meaning that valuation disagreements cannot be resolved by using a discrete set of objective measures. Centocor's bold strategy in partnership negotiations with Schering-Plough demonstrates how a junior partner can resolve a valuation disagreement without compromise by betting on its estimates.

In 1997 Centocor was negotiating a license to Schering-Plough for a Phase II drug, later named Remicade. As the first dedicated treatment for Crohn's disease (prior alternatives were surgery and anti-inflammatory drugs), assessing the market potential was a challenge. Centocor's projections saw the drug growing to $1 billion in international sales by 2007, whereas Schering-Plough's estimates were much lower at $300 million. Confident in their estimates, Centocor asked for a lower percentage on sales up to Schering-Plough's estimates, and a larger percentage if sales exceeded those estimates.[1]

Centocor's strategy worked. The drug exceeded their $1 billion target four years early, achieving $1.2 billion in sales in 2003. Furthermore, the two companies extended the licensing agreement in 2007 and increased Centocor's share of profits.

---

1 Wilan, K. Following the basics enhances forecasting. *Biotech 360*, 2007.

A research-intensive company with limited experience in commercialization may not have sufficient sales and distribution resources to market a product to its fullest potential. In this case, the expense of training and maintaining a sales force large enough to reach an appropriate market may motivate the search for a partner with established channels to market, sell, and distribute the product. International government regulations, language issues, cultural interests, and geographic separation also motivate the selection of complementary partners in serving international markets.

In addition to enabling development and commercialization, partners can also extend institutional support or prestige. Collaboration with an established partner with a reputation for quality products and good service serves as testament to a company's perceived value among industry leaders. Conversely, a single large partner can limit a smaller firm's flexibility by exerting control over research activities and commercialization (e.g., neglecting best-fit applications in favor of larger markets which may be riskier targets), or lead to the perception that a firm has only one customer.

An important consideration in alliance formation is the relative time lines of each partner. Small biotechnology firms may only have sufficient funding to sustain operations for one or two years, giving them a necessarily short-term focus. Established firms must consider longer-term issues and may have a vision extending more than a decade into the future.

Alliances best serve capabilities that are better obtained via partnership than through internal development, outsourcing, or outright acquisition. Because only strategic elements need to be involved and unproductive relationships can be terminated relatively easily, alliances are a less costly, less risky, and more flexible way to acquire capabilities than outright acquisition, especially if the senior partner seeks products and not R&D or sales capacity. Roche chose alliance over acquisition in November 2002 by licensing nearly the entirety of Antisoma's product pipeline in exchange for $43 million in cash and equity in addition to royalties on future marketed products. This transaction granted Roche the drug leads it desired without the need to pay for unwanted elements, and gave Antisoma much-needed cash and continued independence in a poor financing environment.

CONTROL OF ALLIANCES

The issue of control is central to alliances. While alliance parameters and considerations are determined at the outset, developments during the course of an alliance often necessitate modification of agreements. The prototypical biotechnology alliance involves a junior partner responsible for early-stage activities such as research and development and a senior partner that provides financing in return for a share of late stage development or commercialization rights. Each partner is motivated to secure key rights to maximize control and financial returns.

MANAGEMENT OF DEVELOPMENT AND MANUFACTURING

An important consideration in the design and execution of development projects is deciding the target markets and goals of a project. These factors impact many decisions in the course of development. In clinical trials, for example, a key decision is selecting which disease conditions to target. The complex decision of targeting clinical trials may be influenced by a senior partner's desire not to compete with their existing product line, regardless of the junior partner's interests. The partners may also differ in their desire to serve large markets or willingness to pursue high-risk opportunities. A senior partner may be more interested in applications with large markets, for example, than smaller markets with a greater likelihood of clinical trial success. One of the motivations for large up-front alliance payments is to align interests of the junior and senior partner—a senior partner with a greater amount of money invested in a partnership has more to lose from developmental failure.

The role of each partner in early and late stages of development is also important. Because FDA approval extends only to the specific facility where a tested drug was manufactured, a senior partner seeking to produce a drug internally rather than at a junior partner's facilities must undergo FDA approval for the new facility.

INTELLECTUAL PROPERTY

Intellectual property (IP) is one of the key determinants in the valuation of an emergent firm. If both parties are involved in development or independently manufacture the product for sales, the alliance agreement should outline the ownership of IP emerging from the alliance and ownership of IP not directly aligned with the alliance (e.g., tangential findings). Do both parties share the new IP? Is the new IP automatically licensed from the inventor to the other party? Are there restrictions on sub-licensing this new IP to outside parties?

Control of patent litigation is also very important. If a third party infringes on a patent, who is responsible for protecting the IP, and who has authority to decide or decline to protect the IP? While junior partners may lack the resources to defend their intellectual property should a third party infringe, senior partners may lack the motivation to support prosecution. If a technology is licensed exclusively to one party, the licensee will often be responsible for patent litigation, as they have the greatest incentive to protect the patent. Determining responsibility for patent defense can be more difficult when a technology is licensed with non-exclusive terms.

Beyond patents, the transfer of trade secrets is also significant, as this transfer may be necessary to enable senior partners to practice developed inventions. Additionally, publication of research data, which can bolster a junior partner's standing and valuation, can interfere with the ability of either partner to obtain patent protection. Senior partners may elect to review, delay, or suppress publication by junior partners in order to protect their own self-interests.

MARKETING AND SALES

Control of drug marketing confers control over pricing and enables collection of strategic information via a sales force's interaction with consumers. With an established sales force, the senior partner is often best able to manage marketing and sales. Exceptions include niche products and agreements that grant the junior partner rights to specific territories or applications.

Two conflicts that can emerge over marketing and sales are lack of effort, and sales beyond licensed territories and markets. Most agreements include a clause whereby a partner receiving rights to sell a product agrees to use "reasonable efforts" to market and sell the product in the licensed territory or market. The market may be defined by a disease indication or other application or by a geographic territory. The failure of a partner with exclusive sales rights to market and sell a product can deny a junior partner the ability to realize royalties which would have emerged from commercialization. An example is described in the Box *Exubera: When your partner doesn't sell*, where marketing-partner Pfizer declined to market the inhalable insulin product Nektar had spent nearly a decade developing.

Licensing agreements in which partners share marketing and sales rights may divide the rights by application, disease indication, and geographic territory. A common model is for junior partners

---

Box

Exubera: When your partner doesn't sell

Exubera is an inhalable version of insulin that was hailed as a revolution in drug delivery and the most significant innovation in diabetes treatment in decades, and promised multi-billion dollar annual sales (see the Box *Inhalable insulin: Holy grail or tiger trap?* in Chapter 12). The drug was developed by Nektar Therapeutics in partnership with Pfizer, who had exclusive sales rights.

Following weak sales, attributed to a weak marketing effort, Pfizer pulled Exubera in 2007, citing long-term safety concerns, the need for extensive patient training, and patient reluctance to use the cumbersome and conspicuous inhalation device. Perhaps most importantly, Pfizer changed CEOs several times over the course of the Nektar partnership, meaning that internal support for Exubera may have been lost with the associated management changes.

In withdrawing the drug Pfizer paid Nektar a $135 million payment, of which Nektar paid $38 million to its suppliers. Aside from the loss of future revenues, the abrupt partnership termination called Nektar's technology and very business model into question—a challenge for drug delivery companies is that they are often dependent upon partnerships to gain access to drugs to deliver, and they seldom have control over sales.

Nektar acted quickly to respond to the changes. Shortly after the termination of its partnership with Pfizer, Nektar announced the departure of its COO, the elimination of 110 existing jobs and 40 unfilled openings, and a "transition form a drug delivery service provider to a therapeutics drug development organization."

to retain rights to all or select markets in their home territory, and to grant worldwide rights to a senior partner. Geographic divisions tend to be simpler to control, as drugs will often have different names in different markets, making it possible to control license violation by monitoring drug import. Divisions based on disease indications can be more challenging to control because physicians may select to prescribe a product off-label (see section *Off-label use* in Chapter 8) regardless of inter-company licensing agreements.

ALTERATION OF ALLIANCE SCOPE

The unpredictability of biotechnology research extends the need for flexibility in alliances. Protections against poor performance by either alliance member are often included in the form of alteration-of-scope rights. These rights include priority to the funding firm to expand the scope of the alliance through continued development, sales, and sub-license, the ability to cancel the alliance with or without cause, and enabling the junior partner to reclaim licensed technologies if the senior partner's pace of development or marketing is not aggressive enough.

The ultimate means to gain control is through ownership. With significant capital invested in partnerships and future revenues at stake, senior partners may seek equity shares or even controlling interests in junior partners. A secondary effect of granting a share of ownership is that it grants the senior partner an interest in the successful future performance of the junior partner. A downside of having a significant ownership stake is that poor performance or malfeasance by the junior partner may reflect negatively on the senior partner.

MERGERS AND ACQUISITIONS

Biotechnology companies merge to resolve either a lack of size and resources, or a lack of products. For example, companies with innovative technologies and companies with expertise in development and commercialization can mutually benefit by merging to align their respective strengths. While mergers can technically be defined as the combination of controlling interests of two companies, in many cases acquisitions—the appropriation of the controlling interests of one company by another—are politely described as mergers.

The risk of developmental failure in biotechnology necessitates diversification of research efforts. Diversification makes companies less dependent on the performance of individual products and services, and therefore equips them to cope with commercial or developmental failure, competition, and product obsolescence. In the quest for diversification a preferable alternative to developing numerous products and services internally is to merge with, or acquire, another compatible company.

Aside from research synergies and economies of scale, the fusion of two companies can have important implications for funding. Because increased size generally enhances the ability to raise funds, mergers can lead to increased liquidity, attracting investors and enabling a company to make key acquisitions that can further benefit operations.

Combination of research and business elements in a merged company can potentially offer greater-than-additive benefits. Greater bulk can allow merged companies to develop sales, marketing, and manufacturing resources that would not be available to either independent company. The union of compatible companies can provide access to a greater number of markets, the ability to offer customers a range of related products and technologies, and control of key distribution channels. In addition to offering complementary products and services, merging also makes it possible to offer integrated solutions, adding value by easing customers' burden of integration.

To maintain focus it is important that the respective products and services of merging companies share common elements. Ideally, products and services on the market and in development should complement each other and decrease risk with minimal diversification or duplication of efforts. The business models of the companies involved must also be compatible and they should share a similar management philosophy. An example of consolidation without obvious added value is the merger of Procept and Heaven's Door. In January 2000, when public markets were very supportive of Internet companies, 10 year-old biotechnology company Procept and one year-old Heaven's Door merged to produce Heavenlydoor.com, an on-line funeral services firm.

Despite the appeal and advantages of mergers, the challenges of combining management teams and research efforts can frustrate integration. In a merger between similarly-sized firms, the combined entity will likely not be managed by the sum of both company's management teams; each member of either management team may find their position occupied by a member from the opposing company. Unlike "merging" with a larger partner, in a merger of like-sized firms it may be harder for board members to believe that their company is going to be in a better competitive position despite their absence. Institutional investors with significant equity interests and a short-term focus dictated by the type of fund they manage may also resist short-term dilution, regardless of any potential long-term benefits. The Box *Hybritech's crucial role in San Diego's biotechnology industry* in Chapter 10 is a legendary example in poor management team compatibility.

Another impediment is incompatibility of scientific activities. Different labs working on similar projects may use incompatible techniques and tools. Furthermore, because the individual companies will likely have core research efforts as well as some peripheral efforts, a merger based on compatible core research efforts may produce a combined entity with twice as much peripheral research as desired, distracting and excessively diversifying operations. While sales teams may see greater-than-additive benefits from having twice as many salespeople or twice as many products to sell, the same does not hold for research.

Merging offers many benefits, but alternative strategies may be preferable in certain cases. If the goal of a merger is to gain money by consolidating operations or to acquire new technology, there may be better ways to achieve the same goal. Corporate partnerships and licensing agreements allow a company to gain necessary abilities and technologies while avoiding potential negative effects of combining companies.

SPLITTING UP

Spin-offs, described in greater detail in Chapter 10, are the opposite of mergers. The purpose of spin-offs is to enhance the focus on independent elements by placing them in independent companies. This strategy is especially useful when a company consists of elements in different sectors.

Spin-offs may enable a firm to attract funding for disparate projects, and can also separate high-risk from stable-revenue components. By splitting into independent entities, financiers can have a choice of which business they wish to invest in.

A company active in the medical device and drug development sectors, for example, may be a good candidate for division. The desired qualities in a medical device firm are very different from those for a drug development firm. Medical device firms are assessed by more traditional measures—cashflow, revenues, net profits, etc.—while investors in drug development firms tend to be less sensitive to present or even near-term financial measures and instead focus on future potentials.

SELECTING PARTNERSHIP OPTIONS

Licensing, alliances, and mergers and acquisitions offer diverse paths to develop products and enable companies to grow. The selection of which option to employ depends on a number of internal and external factors. Acquisition may be most appropriate for companies with defined prospects such as developed products or technology platforms. Alliances may be more appropriate for companies seeking long-term growth and in need of near-term funding or developmental and commercialization assistance.

An examination of the attribution of product sales to development paths (see Figure 14.2) reveals that most sales were attributed to alliances. One notable benefit of alliances over acquisitions is that they provide additional flexibility. If a partnership dissolves, strong leads can be recovered and developed internally by the junior partner or in partnership with a new senior partner (see the example of Celltech's development of CDP571 in *Dealing With Failure* in Chapter 15). For an acquired firm to continue development of a failed

Performance of development paths

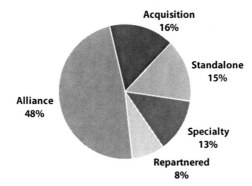

Figure 14.2 *Percent of 2006 product sales attributable to development paths*
Source: Recombinant Capital

lead, they would need to formally spin-off the lead into a new entity. Conversely, acquisition can provide access to substantial development resources. As part of an established firm, the newly acquired entity can access far more resources than would be available in an alliance partnership.

Chapter 15

Managing Biotechnology

> Marketing and innovation are the two chief functions of
> business. You get paid for creating a customer, which is
> marketing. And you get paid for creating a new dimension
> of performance, which is innovation. Everything else is a cost
> center.
> *Peter Drucker*

As described in the opening quotation by the late Peter Druck-
er, innovation and marketing are the two chief functions
of business. The basic role of management, therefore, is to
guide innovation and marketing, and to assemble and maintain the
supportive elements which facilitate them. This chapter integrates
themes from the previous chapters to provide perspectives on managing biotechnology companies.

While biotechnology companies are appropriately described as
R&D-intensive, it is important to recognize that a number of other
operations are required to enable the commercialization and profitable sale of innovations. Figure 15.1 presents a cost analysis of the
pharmaceutical industry as a general model of expenditures in a
mature biotechnology companies. Several of these expenditure categories can be affected by new innovations. Developing products
using biotechnology techniques can directly affect the cost of R&D
and manufacturing—biotechnology drugs tend to cost more to develop and manufacture than pharmaceutical drugs. Secondary effects stemming from the use of biotechnology, such as defining markets through the use of functional genomics or filling unmet market

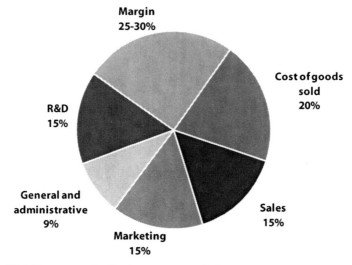

Figure 15.1 *Pharmaceutical industry cost analysis*
Source: Datamonitor, 1999

needs with innovative products, can positively impact marketing and sales operations, reducing costs and increasing profits.

STARTING UP

Biotechnology companies form to develop and exploit new technologies. Conditions in which start-ups thrive are characterized by disruptive changes that permit the development of products and services to satisfy unmet market needs. Biotechnology allows the development of novel products, cheaper and more efficient manufacture of existing products, and development of more effective and refined versions of existing products.

A start-up's goal is to attain sufficient funding to reach milestones that will raise its valuation and lead to either further financing or liquidity. In order to succeed, a company must have a unique competitive advantage that positions it to succeed in its endeavors.

A company's fundamentals change as it matures. While business plan formulation is essential early-on to outline long-term goals, business plans inevitably change as a company matures. Research and development are also very important at early stages, enabling the development of marketable products and a revenue stream, or at

Figure 15.2 *Producing and selling biotechnology products and services*

least the assurance of one.

Once a product has been developed, management needs to establish manufacturing and commercialization abilities. Early marketing and sales agreements provide income and will have a long-term impact on a company's development. To ensure continued success, a balance must be established between research and commercialization; a broad pipeline must be complemented by manufacturing and sales abilities that will result in commercial success.

VIRTUAL COMPANY

Biotechnology companies can benefit from outsourcing elements of their business rather than developing necessary capabilities internally. Selecting partners with specialized expertise and economies of scale to handle certain operations may be cheaper and more effective than employing internal resources and can allow a company to maintain focus on core competencies. An extreme form of outsourcing, the virtual company, outsources all or most business activities. Virtual companies may own or license intellectual property, and can be run by just a few executives who manage the outsourcing activities.

One potential benefit of outsourcing is flexibility. By changing partners, a company can quickly adapt to industry dynamics. This flexibility is balanced by a negative consequence; contract partners likely do not have the same level of dedication and focus as full-time employees. In addition to flexibility, virtual structures also permit greater speed of development. The ability to start and stop independent development activities without the need to develop internal capabilities or restructure internal assets permits just-in-time decision making. For firms engaging in their own research, a partial virtual structure in which most of the elements of development are outsourced allows for a rigorous focus on core competencies.

Two significant concerns for virtual companies are security of intellectual property and control of research and business activities. Outsourcing proprietary techniques and products is accompanied by the risk of losing control of intellectual property. While selective outsourcing allows vital elements to remain secure, absolute outsourcing puts all intellectual assets at risk. Virtual companies may also be at the mercy of contract partner price increases or changes in service quality.

One strategy for effective implementation of a virtual company model is to employ it as a low-cost strategy to perform initial proof-of-principle studies. This defensive strategy can facilitate critical initial studies at a minimal cost, allowing founders to postpone the search for external funding. The advantage of delaying external funding is that a more developed technology will grant a company a relatively higher valuation, permitting founders to receive comparatively more money in exchange for less equity (see Table 11.1 and Figure 17.2). The trade-off of this approach is that important scientific expertise will not be present in-house, requiring repetition of early experiments to develop necessary skills and verify reproducibility.

Many biotechnology companies employ a virtual structure initially, offering academic scientists equity *in lieu* of cash for critical proof-of-principle research. These same scientists are often later installed as scientific directors or advisory board members when the company obtains sufficient funding and adopts a more traditional structure.

MANAGING R&D

A sound management team with proven expertise is crucial for research and commercial success. Management must be able to effectively coordinate successful development and commercial execution. The duration and expense of development mandates a long-term outlook and access to financing at critical stages, necessitating skilled managers. Poor timing can undermine an otherwise promising product. Furthermore, the cost of product manufacture or service delivery must permit profitable sales.

Effective management of innovation, development, and marketing can lead to continued success. In the long term, fostering innovation can permit the development of applications that would otherwise not be possible. However, on an individual project basis it is important to distinguish innovation from market quality. To be commercially successful, a product or service must have customers who are willing and able to pay a profitable price (see Box *Biotechnology myth: Build it and they will come* in Chapter 13). While some products and services may potentially create their own niche, profits are far more certain for solutions serving presently unmet needs. Products such as drugs, for example, should have clear advantages over existing alternatives. They must also address a market of sufficient quality to justify the risk and expense of development.

The high risk of failure of individual research lines in biotechnology necessitates a measure of diversification (see the section *Project Selection* in Chapter 12). While focusing research on a common technique or application enables synergies between individual research projects, it also increases the risk that several projects will fail for a common reason. Conversely, diversification permits companies to engage in multiple mutually exclusive research projects, but introduces the possibility that no project will succeed for lack of ability to overcome significant hurdles or align commercial goals. In order to manage risk, biotechnology companies must develop a balance of focus and diversification.

It is difficult to determine the overall feasibility and financial return of biotechnology projects prior to completion. In the case of drugs, the final outcome from clinical trials is a binary event from which a drug will emerge either as marketable or not marketable. Accordingly, it is essential to continually evaluate research progress, feasibility, and potential outcome in order to effectively manage biotechnology research projects. Regular evaluation of the feasibility of projects, progress to completion, activities of potential competitors, and expected outcomes, enables project managers to objectively assess the quality of, and to effectively manage, multiple research projects.

Box

Entrepreneurial leadership in an established biotechnology company

With its stock up 1,190% in the decade ending June 2006, Genentech, once a legendary startup whose leadership under its founder the late Robert Swanson has been well chronicled, today leads the biomedical industry as an established company. Here are just a few examples of how a strong entrepreneurial culture at Genentech over the last ten years has kept innovation and effective commercialization humming at even a large company:[1]

Entrepreneurial Leadership
- Hiring top scientists. Even as a public company, Genentech has maintained a university-like environment that has attracted top scientists including a former Stanford neurologist and a former University of Washington immunologist.
- Driving scientific initiative through self-selected projects. Genentech encourages its researchers to spend 25% of their time on projects of their choosing (vs. an industry average of 10%). In 1988, Genentech's Napoleone Ferrara focused on anti-angiogenesis, which cuts the blood supply to cancer tumors, resulting in Avastin, a $3 billion (peak sales) colorectal cancer treatment.

Open Technology
- Genentech has in-licensed over 100 technologies such as Rituxan, a $1.6 billion (2004 sales) cancer treatment.

Boundary-less Product Development
- Cross functional product development teams include both technical personnel, relevant business functions, and key outside stakeholders like physicians.
- Using genomics to re-engineer drug discovery. Genentech internally created and then applied in company-wide collaborations the Secreted Protein Discovery Initiative, which generated five product leads. This project accelerated drug lead identification within the human genome by focusing on the 10% of proteins that travel outside cells, blocking or spreading disease.

1 Cohan, P., Unger, B. Four sources of advantage: Technology and bio-medical companies create success cycles. *Business Strategy Review*, Spring 2006.

Disciplined Resource Allocation

- Focusing on areas of therapeutic expertise. Levinson's analysis of the drug industry concluded that greater therapeutic focus yields higher shareholder returns. So when Levinson became CEO in 1995, he focused Genentech on cancer treatment.
- Killing development projects lacking sufficient scientific justification. Levinson kills projects he thinks lack potential. Genentech only moves compounds into clinical trials—which cost between $30 million and $100 million—if the scientific arguments for pursuing the drug can withstand Levinson's intense scrutiny.

The lessons of Genentech are also appropriate to today's biotechnology startups. Indeed a track record of entrepreneurial culture, disciplined resource allocation, and careful technology market-matching[2] is an attraction for venture capitalists funding the expansion of small bio-medical companies. For example, Dan Summa, Venture Partner at Genesys Partners in New York City, emphasizes that he looks for companies that have successfully endured periods of scarce resources. These are entrepreneurial management teams that learned key skills during the post-Internet bubble "nuclear winter" of biotech funding in the early 2000s; skills such as bootstrapping and surviving with leaner infrastructure rather than large organizations, motivating talented people with means other than high salaries and stock options, using in-licensing and partnering to obtain technology from others and to stretch resources, and focusing on well defined revenue-generating projects and business arrangements that demonstrated and advanced their innovative technologies.

Contributed by Dr. Barry Unger (unger@bu.edu) of Boston University and Peter S. Cohan (peter@petercohan.com, http://petercohan.com) of Babson College and Peter S. Cohan & Associates.

---

2 Unger, B., McDonald, I. An exploratory study of mechanisms and processes in the development of U.S. technology-based high growth ventures. Presented at the 4th High Technology Small Firms Conference, 5-6 September 1996. Enschede, The Netherlands.

PORTFOLIO MANAGEMENT

The central challenge of portfolio management in biotechnology is that it is often not possible to predict the outcomes of research efforts. R&D decisions in some application areas, like medical devices, reformulation, repositioning, and combination products, are based on targeting specific market opportunities. When the bulk of innovation is in a new production process or delivery method, R&D can likewise be directed at existing markets. In these cases, the market sizes may be measurable, but the R&D costs may not be known. In many other cases both the cost and outputs of R&D are not known at the outset. This creates a need to regularly evaluate R&D progress and actively direct investments at promising leads.

Cases where R&D outputs can be reasonably estimated but costs are not known are superficially similar to mining or oil exploration, with some important differences. A common method used to manage mining and oil exploration projects is options-based valuation. In these cases it may be possible to assign investment priorities based on projecting future earnings and discounting them based on the perceived risk of failure (see *Valuation* in Chapter 11). As additional information and probabilities are considered, these calculations can become very sophisticated and cumbersome.

While these options-based or discounting models can be effective predictors for ventures such as oil exploration, where revenues, risks, and costs can be reasonably estimated, the unpredictable and dynamic nature of biotechnology product development challenges their application. For example, a drug that is found unsatisfactory for its intended indication may be useful for a second indication. Modeling becomes increasingly complex when secondary and tertiary markets are included. Risk-adjusted valuation methods are more applicable to cases where end products are highly liquid, there is an understanding of the volatility of potential returns, and a large database of comparable situations exists.

The frequent application of human judgement in decision-making, shown in Figure 15.3, echoes the sentiments in the section *Valuation* in Chapter 11. Hard measures for valuation in biotechnology do not exist. However, formal valuation models are essential com-

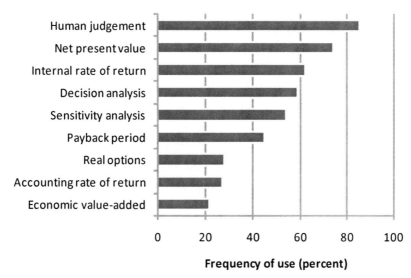

Figure 15.3 R&D *decision-making tools used by small biotechnology companies*
Source: Skrepnek, G.H., Sarnowski, J.J. Decision-making associated with drug
    candidates in the biotechnology research and development (R & D) pipeline.
    Journal of Commercial Biotechnology, 2007. 13(2)99-110.

ponents in supporting value determinations, which require a good
measure of individual judgement based on intuition, experience,
and expertise.

FAIL FAST

A recent study from the Tufts Center for the Study of Drug Development found significant differences between pharmaceutical
companies in the time required to develop drugs.[1] The five fastest
companies between 1994 and 2005—Bayer, AstraZeneca, Allergan,
Boehringer Ingelheim and Merck—had as much as a 17 month speed
advantage over average performers. A follow-up study of clinical development executives at the top-performing firms identified four key
activities associated with increased performance:

- Enterprise-wide adoption of e-clinical technology
  solutions
- High usage of contract clinical service providers

1 Kaitin, K.I. *et al.* Fastest drug developers consistently best peers on key performance
   metrics. *Tufts Center for the Study of Drug Development Impact Report*, 2006. 8:5.

- Active interaction with regulatory agencies
- Effective management and prioritization of resources, including earlier termination of poor projects[2]

The early termination of poor projects was echoed in the Tufts CSDD report, which found that fastest drug developers terminated the majority of cancelled projects in Phase I, versus the slowest drug developers who terminated the majority of cancelled projects in Phase II. This strategy can also deliver great cost savings, as the cost of clinical trials grows progressively with each phase.

INTELLECTUAL PROPERTY PROTECTION

Intellectual property can form a barrier to entry of competitors and attract investors and partners. Because poor security of intellectual property can threaten a company's competitiveness, discourage investors and partners, and put a company's very survival at risk, the quality of a company's intellectual property protection is central to its prospects for success. The sections on *Patent Infringement, Challenge, and Exemptions* and *Extending Patent Protection* in Chapter 7 describe strategies to protect intellectual property and maximize the lifespan of patents.

Intellectual property protection is of central importance in biotechnology. The primary role of intellectual property protection is to enable a company to exclude competitors from a market. While considerable effort is generally required to produce a new biotechnology product, copying is often much simpler. The sophistication and widespread availability of biotechnology research tools limits the ability of trade secrets to protect inventions, due to the likelihood of reverse-engineering by competitors. Because of the importance of maintaining a competitive advantage, the scope of patent protection and market exclusivity have an important influence on which products biotechnology companies will develop. Patents and other forms of intellectual property protection are discussed in greater detail in Chapter 7. In addition to securing future revenue streams, the qual-

---

2 Frantz, S. Study reveals secrets to faster drug development. *Nature Reviews Drug Discovery*, 2006. 5:883.

ity of a company's intellectual property is a key consideration of investors, partners, and acquirers, impacting the value of a company.

MANAGEMENT CHANGES WITH GROWTH

A company's management team is the ultimate source of leadership in strategy, implementation, and financing. In a young company the founding entrepreneur and financial sponsor may play some or all of these roles. As a company grows, it becomes necessary to add specialists and dedicated staff members to relieve the burden on founders, enabling them to focus on the activities for which they are

---

Box

Replacing founders

The management skill sets needed by biotechnology firms change as they mature. A frequent means to gain access to new skill sets is to replace key management. A common point of contention between founders and financiers is when the founders, serving in senior management positions, should step down. In some cases founder-replacement occurs shortly after investment by venture capitalists and in others it may be tied to major liquidity events such as an initial public offering (IPO) or acquisition.

Founders frequently see early replacements as a tactic by financiers to take control of the company they worked passionately to build, and they often question the ability of outsiders to fully understand and properly manage their company. A recent study of 77 venture-backed UK biotechnology firms investigated the correlation of company performance and founder tenure.[1] Companies where the entire founding team continued to play a significant role four years after formation demonstrated a greater number of employees, more products in clinic, more investors, and a higher degree of IPOs and acquisitions. Of all management roles, the tenure of founding CEO correlated most strongly with success measures. Interestingly, post-IPO retention of founder-CEOs was weakly negatively correlated with stock price, suggesting that the best time to replace founder-CEOs is shortly after an IPO.

1 Bains, W. When should you fire the founder? *Journal of Commercial Biotechnology*, 2007. 13:139-149.

best suited and bringing much-needed operational experience to a maturing company.

A difficult time in the growth of any company is the point where founders must step down and allow others to manage the company they have built. In some cases these transitions may be initiated by financiers (see Box *Replacing founders*), and in others they may be initiated by founders seeking to start new ventures, or to free themselves of the burden of running a biotechnology company.

While founders may manage diverse activities such as R&D, partnerships, and human resources, in mature companies these roles require specialized skills, and the people who perform these

---

Box
Flavr Savr tomatoes: Operating in unfamiliar markets

The first genetically engineered food, Calgene's Flavr Savr, was the result of the insertion of a single DNA sequence that interferes with the expression of a gene involved in fruit ripening. Flavr Savr tomatoes stored at room temperature remain fresh long after normal tomatoes rot. This seemingly trivial difference allows greater flexibility in harvesting times and permits farmers to vine-ripen the tomatoes, resulting in a more flavorful product worthy of a higher price.

In the hope that consumers would pay a premium for a better tasting product with a longer shelf life, Flavr Savr tomatoes were introduced to the market in 1994. The tomatoes were removed shortly thereafter, not due to public outcry or even lack of consumer interest, but because of poor commercial execution. For all the development efforts and extensive regulatory testing, Calgene was unable to sell the tomatoes at a profit. If the genetically engineered trait had had a much more dramatic effect in keeping vine-ripened tomatoes firmer for shipping, Calgene might have been able to keep costs down and better meet public demand. Unfortunately, the engineered trait turned out to be of marginal value to the fresh tomato market and Calgene, with minimal experience in that market, could not turn a profit on its novel tomato product. [1]

---

1 For more information, see: Martineau, B. (2001) First Fruit: The Creation of the Flavr Savr tomato and the birth of biotech food. McGraw-Hill, New York.

roles may come from fields unrelated to biotechnology where they have developed their talents. One of the downsides of failing to recruit specialized talent is that the founding management team may be poorly positioned to function effectively in the market in which they operate (see Box *Flavr Savr tomatoes: Operating in unfamiliar markets*). Additional sources of domain expertise, as described in Chapter 10, are the board of directors and scientific advisory board. These esteemed industry leaders and experts can be tapped to fill management gaps in growing firms.

DEALING WITH FAILURE

Guiding products through development can be tricky. Ideally, development paths with a low probability of success should be eliminated as quickly as possible. Late-stage failures or evidence that a key project's development path is unlikely to yield success may require drastic measures to sustain investor confidence and find new paths to success.

Companies that experience significant failures have several options. These options include renaming the company, shuttering it,

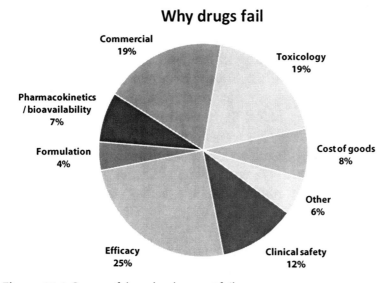

Figure 15.4 *Causes of drug development failure*
Source: Kola, I., Landis, J. Can the pharmaceutical industry reduce attrition rates? *Nature Reviews Drug Discovery*, 2004. 3:711-715.

merging with another firm, changing management, or refocusing development efforts. Internal factors, such as a company's cash position and remaining development opportunities, and external factors, such as the esteem of management among potential and current investors, influence which strategy may be preferable.

CHANGING NAMES

Changing names is a relatively straightforward way to distance a company from past experiences. While the immediate impact of a name change is unlikely to restore faith among disillusioned investors, it may be effective as part of a long-term strategy for recovery.

Another purpose of changing company names is to take advantage of market trends or to reinforce company fundamentals. In June 2000, when genomics firms were favored in public markets, Incyte Pharmaceuticals changed its name to Incyte Genomics to reflect the company's approach to drug discovery. To reflect a corporate restructuring, PE Corporation changed its name to Applera, combining the names of the company's two operating businesses, the Applied Biosystems Group and the Celera Genomics Group.

CHANGING MANAGEMENT

Failure of key products can prompt investors to demand management changes. The high-profile failure of ImClone Systems' promising anti-cancer drug Erbitux in late 2001 and the precipitous fall of ImClone's stock amid allegations of poor trial design and insider trading led to immediate management changes (see Box *Erbitux: Poor study design* in Chapter 8). In another example, after sequencing the human genome Celera found that selling genomic sequence information offered insufficient revenue growth potential. Celera responded by selecting a new CEO and replacing numerous upper-level executives to reconfigure itself as a drug development firm.

TARGETING NEW MARKETS

Biotechnology product development is extremely challenging. Setbacks may occur at any point in development and may require modification of techniques, product reformulation, or new development strategies.

For drugs that fail early in development, often due to poor safety profiles, there is little motivation or incentive to continue development. Alternatively, drugs that are safe but are not effective for their initial therapeutic target may still be useful for other treatments.

Box
Preventing a product recall from bankrupting a company

Merck voluntarily withdrew its anti-inflammatory drug Vioxx from the market in September 2004 in response to evidence that the drug could be responsible for heart complications seen in long-term use. The Vioxx saga is often cited as a case study of the dangers of the blockbuster model, as the large population which helped Vioxx deliver great revenues quickly became a liability. 27,000 lawsuits covering 47,000 claimants were filed against Merck, alleging that Merck failed to adequately warn patients and intentionally hid data on dangerous side effects. Estimates of Merck's potential tort exposure measured in the tens of billions of dollars—enough to potentially bankrupt the firm.

Merck's strategy to manage these lawsuits was to prevent class-action suits at all costs. While victory in class-action suits could potentially settle many cases at once, the cost of losses could be devastating. Merck set aside $1.9 billion for legal fees and elected to handle each case individually. A risk inherent in class-action suits is that undeserving plaintiffs may join the lawsuit and inflate damage awards. By insisting on case-by-case litigation, Merck was able to expose weaknesses in many of the claims and set themselves up for a more favorable settlement.

After sixteen trials (Merck won eleven and lost five) and legal fees exceeding $1 billion, Merck offered a $4.85 billion settlement to plaintiffs who met certain criteria. Plaintiffs who Merck felt had weaker cases were excluded from the settlement. While Merck's tenacious tactics and large settlement offer may prove effective, this strategy doesn't always work. American Home Products, now part of Wyeth, set aside $21.1 billion for Fen-Phen litigation (see Box *Fen-Phen: Risks of off-label use* in Chapter 8), and their settlement is in its seventh revision.[1]

---

1 Fisher, D. Will the Vioxx settlement work? *Forbes*, November 13, 2007.

Minoxidil and Viagra are two popular drugs that were initially investigated for vascular conditions and were eventually marketed for hair loss and impotence, respectively. In another example, Celltech developed an antibody named CDP571 to treat septic shock and licensed it to Bayer. When Bayer dropped development of the drug, Celltech's shares fell 46 percent in one day. Persevering, Celltech ultimately managed to develop CDP571 for the treatment of Crohn's disease.

CHANGING BUSINESS MODELS

There are often several ways in which a biotechnology company can commercialize its intellectual property (see Table 17.1 in Chapter 17). When companies with strong fundamentals face significant setbacks, they may elect to change business models to capitalize on their strong cash positions or robust intellectual property assets. For example, after Pfizer withdrew Exubera, the inhalable insulin drug developed by Nektar, from the market, Nektar faced a crisis (see Box *Exubera: When your partner doesn't sell* in Chapter 14). Deprived of future royalty opportunities, and with other potential partners likely questioning the value of Nektar's drug delivery technologies, the company elected to change their business model from supplying delivery technologies to others, to developing novel therapeutics internally.

LIQUIDATING ASSETS

Biotechnology companies that are short of financial resources and have poor prospects of finding a buyer or raising sufficient resources to permit continued development may find liquidation of commercial assets and return of any proceeds to shareholders the only remaining strategy. The value of intellectual property in biotechnology often means that while a company may cease to exist, other firms may acquire promising technologies and continue development.

COMMUNICATIONS AND PUBLIC RELATIONS

Biotechnology companies must communicate with a variety of audiences, including investors, current and potential future collaborators, prospective employees, and the media. It is important to consider the different needs of each of these distinct audiences in managing public relations.

INVESTORS

Private and public biotechnology companies need to maintain the support of investors to ensure continued financing of business operations. In addition to direct company communications and press releases, investors may also use additional sources to gain corporate information, including trade and financial magazines and newspapers, independent websites, analyst reports, and conferences, necessitating the effective management of reports emerging from these outlets.

Investors must be provided assurances that their investment will deliver an acceptable return; evidence of developmental progress; adherence to corporate strategy; and, maintenance of operational stability. Because publicly held companies have a responsibility to promptly publish any information that has the potential to affect share prices, the requirement for prompt notification of significant events and developments should be an important consideration in deciding when to take a company public.

Institutional investors and professional investors may own significant portions of equity in public and private companies. Their large equity stakes make these investors relatively more interested in the future success of a company and grants them a greater level of authority than smaller holders, encouraging them to be relatively more active in shaping company policy and actions (see Box *Shareholder activism* in Chapter 10). The purchase and sale of publicly held shares by a large investor may also be interpreted by the markets as a judgment of the merits of a company, potentially triggering large fluctuations in share price. Accordingly, communicating with large equity holders extends well beyond basic public relations.

A very important communication stage for biotechnology companies is the period surrounding an initial public offering (IPO), as early perceptions and investment decisions by analysts can have long-term impacts on share prices. A study of the 57 biotechnology IPOs listed on NASDAQ between 1997 and 2002 found that messages contained in the IPO prospectus had a strong bearing on post-IPO performance. Specifically, consistent presentation as a "prospector" (targeting broad markets and continually searching for new market opportunities) was negatively correlated with 30-day initial returns, whereas consistent presentation as a "defender" (targeting narrow product/market domains and defending them aggressively) was positively correlated with 30-day initial returns.[3]

PARTNERS

Because biotechnology companies rely on partnerships and collaborations to enable growth, it is important to effectively communicate strengths to potential partners. Partners are more demanding than investors and seek evidence of complementary expertise and products in addition to indications of strong corporate fundamentals and growth.

Some important qualitative differences between investors and partners mandate specially-crafted messages. Whereas investors may focus their activities in certain sectors, their primary goal is to derive a profit on their investment. Partners must also consider compatibility with their business objectives and the potential for a partner's poor performance or misconduct to reflect poorly on their business. For example, Bristol-Myers Squibb faced scrutiny following the improper clinical trial design and insider trading saga at ImClone (see *Erbitux: Poor study design* in Chapter 8) due to their $2 billion in alliance and equity investments in ImClone. ImClone's poor clinical trial performance and poor display of corporate governance led investors to question Bristol-Myers Squibb's ability to vet the scientific and corporate qualities of partners.

A common outlet to communicate with existing and prospective investors and partners is peer-reviewed scientific journals. Scientists publish in scientific journals as an integral component of their ca-

3 Gao, H., Darroch, J., Mather, D., MacGregor, A. Signaling corporate strategy in IPO communication. *Journal of Business Communication*, 2008. 45(1):3-30

reer development, but these media outlets can be very useful for biotechnology companies as well. Publication in an esteemed journal can demonstrate that a biotechnology company's core technology or research findings have passed rigorous scrutiny by recognized experts, facilitating due diligence for investors and partners.

It is also important to maintain a consistent corporate image. The presence of multiple companies with similar objectives and methodologies may confuse partners. If a company cannot successfully convey its uniqueness, it risks being labeled as a company with poor fundamentals continually changing its image to align itself with market trends. Individual companies must differentiate themselves from competitors by conveying the innovativeness and effectiveness of their tools and technologies.

Another compelling reason for a company to differentiate itself from competitors is to provide a shield from poor performance of competitors. When multiple companies have similar products in development or on the market, a developmental failure or market withdrawal by one company can negatively impact the others. Entire application areas have been sidelined by the poor performance of individual trials. Certain alternative financing vehicles have likewise been abandoned due to scandals by companies outside the biotechnology industry. It is therefore important to monitor the performance of competitors and be prepared to help investors and partners understand important differences to prevent the spread of negative perceptions.

EMPLOYEES

The challenges faced in attracting employees are similar to those faced in attracting partners. Employees seek employers who are aligned with their goals and can help in their career progression. Relatively unknown companies and those with poor reputations may have difficulty recruiting and retaining quality staff. An important element in attracting new employees and motivating and retaining existing employees is reinforcing the notion that a company is (or will be) prestigious and distinguished in its field. This message can be conveyed by ensuring frequent mention in media outlets, maintaining a steady publication stream, and actively participating in trade events such as conferences.

MEDIA, CONSUMERS, AND GENERAL PUBLIC

While investors, partners, and employees develop their perceptions of biotechnology companies through corporate reports, publications, and trade shows, it is also important to convey a positive impression to the general public. Public opinion can motivate political decisions such as price controls and technology and product bans that can have a profound influence on a company's operations.

Because the media can influence the opinion of investors, partners, employees, and the general public, it is vitally important to ensure effective communication with the media. This is communication is challenged by the need to ensure comprehension by a lay audience. An Italian study of consumer perceptions found that increased exposure to coverage of science in the media does not lead to increased understanding of science and may in fact lead to reinforcement of incorrect notions, emphasizing the need for management of communication.[4] Because overly technical or obscure messages may confuse the public, it is imperative that biotechnology companies promote a positive understanding of their goals in public messages by ensuring that common themes and messages are present in all forms of communication.

---

4 Bucchi, M. Neresini, F. Biotech remains unloved by the more informed. *Nature*, 2002. 146:261.

Chapter 16

International Biotechnology

Cuba's future must, by necessity, be a future of scientists.
Fidel Castro, shortly after the Cuban revolution.

Biotechnology is a great driver of economic growth, and this opportunity has not been lost on governments around the world. Many countries are taking aggressive steps to support the growth of local biotechnology industries in an effort to reap the benefits of having solutions developed for local needs, to leverage existing science and technology infrastructure, and to realize increased tax revenues from increased numbers of high-paying jobs and global sales of high-value products.

A prime example of the value of investments in biotechnology is the growth of Cuba's industry. Early efforts in Cuba were focused on producing generic versions of drugs approved elsewhere, as a means to address local needs. They also engaged in partnerships with foreign companies, offering low-cost R&D opportunities while expanding domestic research infrastructure and expertise. Leveraging this strong base, combined with exceptionally strong and sustained governmental support, Cuba has produced several novel drugs for diseases endemic to developing nations and has produced more innovations than many more-developed nations.[1]

Countries around the world differ in their laws, regulations, and economies. This chapter describes how these factors impact biotechnology development and presents some strategies for international

---

1 Mola, E.L., Silva, R., Acevedo, B., Buxadó, J.A., Aguilera, A., Herrera, L.
Biotechnology in Cuba: 20 years of scientific, social and economic progress. *Journal of Commercial Biotechnology*, 2006. 13(1):1-11.

Box
Why did the biotechnology industry start, and prosper, in the United States?

It is often taken for granted that the United States is where biotechnology was born, and is home to most of the successful companies in the industry. The unanswered question, however, is how this level of success came to be. Examining the factors contributing to the early development of biotechnology, and to the continued success in commercializing biotechnology, yields insights on how to succeed in biotechnology.

The key scientific discoveries which laid the technological foundation for the biotechnology industry are described in the section *Knowledge and Skills* in Chapter 2. Most of the individuals who received Nobel Prizes for the key discoveries were American. The common national origin of these scientific leaders suggests that a) the United States likely had many other prolific scientists, and b) there were ample opportunities for local collaboration, facilitating spill-over from the laboratories of these Nobel laureates and other profilific scientists.

To answer the question of how the United States was able to cultivate this group of outstanding scientists, one can look at R&D funding. According to figures from the Organisation for Economic Co-operation and Development (OECD), in 2006 the United States accounted for 42 percent of R&D funding by OECD member nations. This leadership has persisted for decades; in 1973, when Stanley Cohen and Herbert Boyer developed methods for gene splicing, the United States' share of OECD R&D funding was 55 percent, meaning that the United States spent more on R&D than all other OECD members combined.

While the United States currently has many supportive regulatory policies and opportunities for funding, this was not always the case. Shortly after Cohen and Boyer's demonstration of gene splicing, the technology was banned by an international moratorium. The popular Bayh-Dole Act and SBIR funding program were not implemented until the first biotechnology companies were well on their way to commercialization. Therefore, while supportive policies and funding opportunities play an important role in the biotechnology industry today, other elements were necessary in their absence in the industry's initial years.

In the absence of federal funding for applied research, early biotechnology companies turned to another source: venture capital. The United States venture capital industry grew in concert with the semiconductor industry, and success in that sector demonstrated to financiers and biotechnology scientists alike the opportunity to realize significant financial, technological, and societal gains from investments in emerging technologies.

Industry partners and receptive public markets also played an important role in supporting the nascent biotechnology industry. Without sufficient resources to independently develop, manufacture, market, and sell drugs, biotechnology companies needed industry partners and the opportunity for investors to realize a near-term return on their investments (as opposed to deriving a return from long-term revenues). The American pharmaceutical industry, facing weakening pipelines, became ready partners to co-develop biotechnology products, to acquire drug leads, and even to purchase biotechnology companies, enabling investors to realize positive returns on their investments in advance of product commercialization. The public stock markets were also very receptive to biotechnology companies, enabling them to raise sufficient financing well in advance of profitability.

Adapted from: Friedman, Y., Seline, R.S. "Innovation in the United States and Germany: Case studies." *American Institute for Contemporary German Studies.* Washington DC, 2007.

R&D and commercialization.

BENCHMARKING AGAINST THE UNITED STATES

Individuals in many countries benchmark their local biotechnology industry against the U.S., or bemoan the relative challenge of raising capital or attracting top talent domestically. Some also recommend domestic innovation reforms to match current U.S. laws and incentives. Such comparisons and development plans need to be viewed in perspective to the size of the U.S. economy and the distribution and history of the U.S. biotechnology industry.

The U.S. is home to the world's largest economy and the world's

Biotechnology companies

Figure 16.1 *Distribution of U.S. biotechnology companies*
Source: Beyond Borders: The Global Biotechnology Report 2006. *Ernst & Young*, 2006

largest pharmaceutical market, and it does not have price controls. The relatively consistent legal and political frameworks across states facilitate early-stage investments, mergers and acquisitions, and market access. Multinational companies also tend to locate their operations closest to their largest markets, which is a strong continuing driver for U.S. growth. It is therefore unfair to expect a country with a smaller economy to have a comparative-sized biotechnology industry.

Furthermore, the U.S. biotechnology industry is not evenly distributed geographically (see Figure 16.1). While there are several very successful regional concentrations, a majority of U.S. states are struggling to build their local biotechnology industries—a challenge shared by many countries. Given that all states adhere to the same basic federal laws supporting biotechnology (e.g., Bayh-Dole Act, SBIR funding, patent laws, etc.) but differ in the size of their biotechnology industries, the solution to developing a local biotechnology industry cannot simply be to adopt U.S. federal laws and incentives; local conditions play a significant role. Additionally, the current set of U.S. federal regulations and incentives helped the biotechnology industry grow to its current state, but do not necessarily translate to

Public companies

Revenues ($mm)

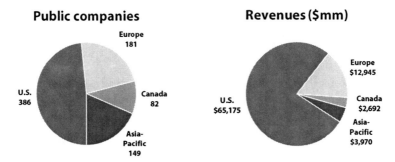

Figure 16.2 *Global distribution of public biotechnology companies and revenues*
Source: Beyond borders: Global biotechnology report 2008. *Ernst & Young,* 2008

different political systems and economies.

INTELLECTUAL PROPERTY PROTECTION

The scope of patentable inventions varies among countries.[2] For example, inventions in the U.S. must be *novel, non-obvious,* and *useful,* whereas European patent law requires that inventions demonstrate *novelty,* an *inventive step,* and *industrial application.* Subtle differences in terminology can have significant implications in the scope of patentable inventions, and are compounded by numerous additional restrictions (such as Canada's restriction on patenting higher life forms) and rules for patent filing.

The U.S. is unique in granting business model patents. Additionally, some countries do not offer protection for pharmaceutical products but only allow patents on production methods. The requirements for utility vary in stringency in different countries as well. The U.S. also differs from many other countries in granting a one-year grace period for patent application following initial public disclosure.

A country's IP protection laws have a significant impact on local innovation-based industries, such as biotechnology. Many developing nations have intentionally weak IP protection. This is largely viewed as a strategy to reduce the cost of manufacturing goods and to help support the growth of domestic industries by enabling them

2 For a topical review, see: Kowalski, T.J., Maschio, A., Megerditchian, S.H. Dominating global intellectual property: Overview of patentability in the USA, Europe and Japan. *Journal of Commercial Biotechnology.* 2003. 9(4):305-311.

to manufacture generic versions of goods using technologies patented elsewhere. But, this support for local industry comes at a cost: poor IP protection also discourages inward investment by foreign firms.

In January 2005 India amended its patent laws to recognize product patents, largely complying with the World Trade Organization's Trade Related Aspects of Intellectual Property (WTO-TRIPS) agreement (see Box *Novartis cancels Indian investments over patent dispute* for details on an important exception). Some feel that the early period of weak IP protection was necessary for India to build a strong base of generic pharmaceutical firms—well-positioned to grow into innovative leaders—while others argue that the years of weak patent protection merely discouraged inward investment and delayed the maturation of India's pharmaceutical sector and production of innovations directed at the needs of Indians.

Box
Novartis cancels Indian investments over patent dispute

In January 2005 India amended its patent laws to recognize product patents. A notable deviation from international norms, as described in the World Trade Organization's Trade Related Aspects of Intellectual Property (WTO-TRIPS) agreement, is that relative to known substances, new drug products must "differ significantly in properties with regard to efficacy." This exclusion, which isn't applied in other TRIPS-compliant countries or to other technologies in India, is designed to prevent "evergreening" by patenting superfluous incremental innovations. It also risks inhibiting substantial incremental innovations.

In August 2007, after Novartis failed to receive patent protection for its cancer drug Glivec (sold as Gleevec in the U.S.) in India, the company responded by announcing the elimination of plans to invest hundreds of millions of dollars in R&D facilities in India. It is worth noting that Novartis provides Gleevec free to most patients in India. Claiming that the unfavorable patent ruling was "not an invitation to invest in Indian research and development," Novartis' CEO stated that the company was going to redirect its investments to countries where it has greater intellectual property protection.

Patent protection in more than one country requires filing patent applications in multiple countries (or, when relevant, regional patent offices such as the European Patent Office and others), and is complicated by differences in enforceable claims. The Paris Convention for the Protection of Industrial Property (March 20th 1883, amended on September 28th 1979), signed by most industrialized nations, permits an inventor in a signatory country to file an application first in his/her home country, then file corresponding applications within one year in other signatory countries. The effective filing date for the secondary applications, however, is the date the patent was filed in the original country. This can be a significant element in prior art (information demonstrating that a patent is not novel, and is therefore invalid) cases as any publication before secondary application dates, but after original application, is not considered prior art.

While prior art rules require that the patent filer be the first to practice an invention in the United States, it is possible to patent an invention in the United States within one year of its first public use by another party, provided the initial use occurs outside the U.S. Priority to the first to invent is also unique to the U.S. Most other countries grant patent rights to the first to file. The implication of this distinction is that whereas laboratory notebooks can be used to invalidate a patent in the U.S. by demonstrating prior invention, such protests do not apply elsewhere. The obligation to disclose the best mode to practice an invention is also unique to the U.S.

REGULATION

Differences in regulatory requirements between countries can have a profound influence on the types of drugs approved and on the availability of new drugs approved elsewhere. For most countries there is a single regulatory body, such as the FDA in the U.S., but in the EU submission to the European Medicines Agency can grant approval in all EU member countries.

Regulatory policies can also dampen innovation. Japan's regulatory policies are blamed for a lag in drug approvals relative to other countries. While Japan is the world's second largest pharmaceutical market (the U.S. is the largest market, and the aggregate of the EU is also larger), fewer new drugs are available in Japan, and Japanese approvals generally occur years after approval in other countries. A recent study of biopharmaceutical approvals in the U.S., EU, and Japan found that 91 percent of biopharmaceuticals approved between 1999 and 2006 were approved in the U.S., 80 percent were approved in the EU, and only 34 percent were approved in Japan. Furthermore, while the mean approval lags in the U.S. and EU were 3.7 months and 7.5 months, respectively, the mean approval lag in Japan was 52.6 months—more than four years.[3] Among the factors blamed for this lag is the stringent requirement for Japanese drug applications to cite clinical trials using Japanese subjects, which motivates sponsors to initiate Japanese trials only after trials in other countries have shown promise, delaying submission of drug applications.

LEVERAGING FOREIGN APPROVAL

Foreign approvals can be used to support domestic approvals in several ways. Approval timelines and costs vary from country to country, so it may be advantageous to seek initial approval in a foreign country to speed time to market or reduce development costs. Data from drug administration in foreign markets can aid in clinical trial design, helping select target indications and speeding approval. Information on foreign market size and characteristics can also help establish market size estimates, facilitating financing. Additionally, sales in foreign markets can help fund domestic operations (see the section *Generating Sales Before Marketing Authorization* in Chapter 13).

MEDICAL TOURISM

Medical tourism is the practice of travelling to another country for medical treatment. Motivations include saving money, accessing treatments not available domestically due to long waiting lists, high cost, or lack of domestic regulatory approval.

While medical tourism can be an effective way for patients to

---

3 Tsuji, K., Tsutani, T. Approval of new biopharmaceuticals 1999–2006: Comparison of the US, EU and Japan situations. *European Journal of Pharmaceutics & Biopharmaceutics*, 2008. 68(3): 496-502.

access medical interventions for lower costs and with shorter waits than are available domestically, there are several significant downsides. Saving money on treatments may come at the cost of regulatory oversight, support for complications, or patient rights. For example, the ability to obtain compensation for poor performance or malpractice may be limited, potentially outweighing any other benefits. Far more serious, however, is the present possibility to access procedures which have not received regulatory clearance from any international bodies. For example, it is currently possible for diabetics to obtain pig islet cell transplants and for heart patients to obtain stem cell transplants in certain countries. While such treatments may be viable, and may be the only option for some patients, they may also lack sufficient regulatory oversight and carry the potential to introduce new human diseases.

OPERATIONS

Motivations to perform operations in foreign countries include addressing capacity bottlenecks, accessing specialized talent or expanded workforces, increasing market access, and saving money.

Differences in time zones, measurement standards, and general communication issues can complicate collaborations and slow progress in international partnerships. Call centers, for example, are an example of a business operation which promised great cost-savings, but where the overall cost of implementation and decreased service quality failed to meet initial expectations.

In addition to the intellectual property protection and filing issues described earlier in this chapter, the risk of losing control of IP is associated even with domestic outsourcing. An added risk encountered in outsourcing proprietary elements to countries with lax IP protection is that it may not be possible to prevent unauthorized use. A strategy to prevent loss of IP is to outsource only non-vital components or to distribute sequential IP elements to several partners.

RESEARCH AND DEVELOPMENT

The primary motivation for a company to conduct R&D in a foreign country is to save money. Favorable taxation, described later in this chapter, can reduce costs, as can lower wage and operational costs. It is estimated that in China, for example, drug research costs are 20 percent of those in Western countries.

An important consideration in performing R&D overseas is the protection of patented methods used in R&D, and the potential outputs. Performing certain research activities offshore can also be an effective strategy to avoid licensing technologies which are patented domestically. The Box *When research is done abroad* illustrates how Bayer was able to use Housey Pharmaceuticals' patented drug screening methods overseas and then import information describing the resulting drug leads. In ruling on the case, the Federal Circuit deter-

---

Box

When research is done abroad

The 2003 case of *Bayer AG and Bayer Corporation v. Housey Pharmaceuticals, Inc.* underscores the importance of international patent protection. In this case, Housey sued Bayer alleging that Bayer used their patented drug screening methods overseas and imported information describing the resulting drug leads. Bayer countersued, challenging that Housey's patents were invalid, unenforceable, and not infringed.

The arguments in the case did not focus on whether Bayer had actually used the processes, but rather addressed whether import of information produced by the processes constituted patent infringement. While U.S. patent law enables inventors to prevent others from importing, using, or selling products made by patented methods, the Federal Circuit determined that information describing drug leads is not a "product," thereby allowing its import. Had the patented processes described processes directly involved in the final manufacture of drug products, Bayer likely would have been found guilty of infringement.

This ruling has broad implications for drug discovery and bioinformatics patents that produce information rather than physical products, essentially rendering their outputs unprotectable.

Clinical trial locations

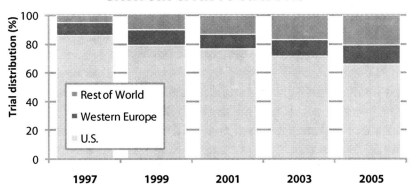

Figure 16.3 *Globalization of clinical trials for FDA approval*
Source: Outlook 2007. *Tufts Center for Drug Development*, 2007

mined that information describing drug leads is not a "product," and therefore allowed its import.

CLINICAL TRIALS

A common concern in performing foreign clinical trials is the treatment and safety of trial subjects. The FDA has specific requirements to qualify data from foreign clinical trials, ensuring that trials must meet the same requirements as those conducted in the United States, and must be conducted in accordance with the laws and regulations of the country in which the research was conducted.

It is estimated that clinical trials in China can cost as little as 15 percent of those in Western countries[4] and those in India can cost 40 to 50 percent as much[5], but this is not the only motivation to perform international clinical trials. Another important factor driving international clinical trials is speed of trials. International clinical trials provide an opportunity to access a larger population, and patients who may have also never received alternate treatments, which can speed recruitment and produce more compelling data. Because patents are filed in advance of clinical trials, reducing the duration of clinical trials can mean more time for patent-protected sales.

---

4 The rush to test drugs in China. *BusinessWeek*, May 2007.

5 Fee, R. A passage to India. *Drug Discovery & Development*, March 1, 2007.

MANUFACTURING

In 2007 AstraZeneca announced plans to eliminate internal manufacturing activities and largely outsource them to countries like India and China. While the actual extent of AstraZeneca's international manufacturing plans have yet to be seen, companies have been increasingly sending manufacturing operations overseas. Because manufacturing is a recurring expense, unlike R&D expenses which are restricted to pre-commercial stages, any cost-savings can have long-term impacts. Pre-clinical manufacturing is also a good candidate for international outsourcing, as it can save start-up firms precious cash.

Just as certain patented research steps can be performed in countries where they are not patented, and the research results may be imported without infringing on domestic patents, it may also be possible to use patented manufacturing steps in foreign countries and to import selected products without penalty. The key to importing the product of patented process is that the product being imported must be materially different from the end-product of the patented process.[6]

A significant risk involved with offshoring any elements in the drug supply chain to foreign countries is the lost ability to track the safety and integrity of products. Drugs manufactured and distributed within a single country can be tracked from source to patient by domestic regulatory authorities. When the supply chain is international, tracking drugs and ensuring quality and integrity at all levels becomes far more complicated. An example of this loss of control is the 2008 contamination of Baxter's heparin drug, which was implicated in dozens of deaths and allergic reactions in hundreds of patients. In an apparent case of economic fraud, an FDA inspection found that the batches of Baxter's heparin contained large quantities of a modified version of a dietary supplement which can mimic heparin in chemical tests, but is far less expensive to produce. The drug was produced for Baxter by a U.S.-based partner who sub-licensed manufacturing to a Chinese firm that had never been inspected by

6 Familant, S.B. Kinik: Raising the stakes for importing products derived from US patented processes practised abroad. *Journal of Commercial Biotechnology*, 2005. 11(4):364-368.

the FDA. While accountability for the drug contamination has not been assigned, this case illustrates the challenges of outsourcing manufacturing beyond the reach of domestic regulatory authorities.

MARKETING

International marketing, like many other international activities, can be very complicated. Different markets have different regulations, and the drivers of sales can also vary. For example, the ability to advertise direct-to-consumers or to brand generic drugs can impact marketing strategies. One solution to this fragmentation is to license global sales to specialists. Many U.S.-based companies retain domestic rights and license global rights in whole or in parts to foreign or multinational pharmaceutical companies.

Drugs are often sold for different prices in different countries due to government price controls or pricing regimens established by drug companies. This creates an opportunity for parallel trade, where consumers in countries where drug prices are higher import drugs from countries where prices are lower. While this strategy can enable individual consumers to obtain drugs for reduced prices, it has been met with resistance from drug companies who wish to independently control prices. One strategy used to prevent parallel trade is to limit supplies of drugs to countries where prices are low, providing them with sufficient product for domestic use but not enough for export.

Along with parallel trade comes the risk of counterfeit drugs. Beyond the potential for revenue loss due to domestic customers buying drugs at foreign price-controlled rates or from unlicensed drug manufacture is the threat that the drugs may be counterfeit and may not contain any active compound or, worse, may contain toxic compounds. For example, drugs purchased from foreign pharmacies, even those in 'trusted' nations, may come from suppliers in third-countries with little or no regulatory oversight.

FINANCE

INVESTING AND PUBLIC MARKETS

The U.S. is home to the world's largest biotechnology market and the greatest number of public biotechnology companies. This strong representation makes U.S. stock markets desirable for global companies seeking the stock price stability and liquidity benefits associated with selling to investors familiar with biotechnology in a market with a substantial quantity of peers.

The relative ease of biotechnology companies in any U.S. state to list on NASDAQ, compared with the hurdles of international transactions in inter-EU country listings also promotes U.S. stock markets and is a hurdle for EU growth. Whereas investments between U.S. states are relatively simple, some EU country policies are at odds with investing in innovative sectors such as biotechnology, and even investments between hospitable countries can be burdensome.[7]

There are also factors discouraging listing on U.S. public markets. For example, expensive Sarbanes-Oxley accounting and auditing structures are an onerous burden for small biotechnology firms which are typically cash-poor. The focus on relatively more mature companies in larger American public markets also encourages smaller firms to list on foreign or smaller markets, where the expectations of investors may be lower. A potential downside to listing on foreign and smaller markets is that companies may not enjoy the full benefits of follow-on financing and stock price growth available in large markets like NASDAQ. However, this strategy may be a good intermediate step for eventual listing on preferred markets.

TAXES

A strategy to decrease a company's tax burden is to restructure in a corporate inversion. Because U.S. corporations must pay federal taxes on domestic *and* international sales, a company can reduce its tax burden by forming a parent corporation in a tax haven country—a country with little or no corporation tax—and forming a U.S.

---

7 For a topical review, see: Removing obstacles to cross-border investments by venture capital funds. *European Union Directorate-General for Enterprise and Industry*, 2007.

> **Box**
> ## Merck's Bermuda subsidiary runs afoul of U.S. tax laws
>
> In 1993 Merck set up a Bermudan subsidiary and transferred patents covering two of its drugs to this entity. Merck then paid the subsidiary licensing fees for these patents in a tax-reduction strategy known as earnings stripping.
>
> It is estimated that Merck avoided $1.5 billion in taxes over the next ten years by using these license payments as deductions against U.S. taxes. The Internal Revenue Service challenged Merck, claiming that the Bermuda subsidiary did not have "independent economic substance"—U.S. companies cannot reorganize for the sole purpose of avoiding taxes. Merck argued that the Bermuda corporation was established to hold the patents as collateral for an investment by a British bank, and was therefore a legitimate entity. This argument was apparently not satisfactory, as the IRS announced in February 2008 that it had reached an agreement with Merck resulting in a payment of approximately $2.3 billion in federal tax, interest, and penalties; one of the largest tax evasion settlements ever imposed by the IRS.

subsidiary to manage U.S. sales. This structure enables foreign sales to be taxed in local markets only, and not be taxed in the U.S. An additional benefit may emerge from "earnings strippings," where the payments from the U.S. subsidiary to the foreign parent for loans, royalties, licenses, etc., are used as deductions against U.S. taxes.

Tax arbitrage is a variation on corporate inversion. Because different countries have different tax laws, it may be favorable to locate specific operations in countries with favorable tax treatment for those operations. Puerto Rico, for example, offered tax exemptions for certain manufacturing activities by U.S. businesses, enabling it to grow to become the largest single source of pharmaceuticals in the U.S. market.

Countries, states, and regions seeking to develop innovative industries frequently offer incentives for qualifying companies as well. For example, France, established the Young Innovative Company (jeune enterprise innovante) program in 2004 to reduce the relatively high tax burdens of starting innovative ventures. Qualifying companies must have fewer than 250 employees, less than €40 million in revenues or less than €27 million in total assets, be less than 8 years old, and must spend at least 15% of their total annual expenditures in R&D. Under the YIC program, companies may receive total corporate income tax exemption for the first three profitable years and 50 percent relief for the following two years, up to €200,000. Investors in YIC companies also receive an uncapped capital gains tax exemption if the shares are held for at least three years. 1,700 french companies enrolled in the YIC program in its first three years, approximately 20 percent of which were biotechnology firms.[8]

Taxes can also stifle innovation. In a recent global survey of venture capital investors, Deloitte Consulting found that Canada was most-cited as an unfavorable tax area. The perceived impact of these sentiments is that while investors may make early investments in Canadian companies, the companies must often move or form headquarters in other markets—usually the U.S.—to obtain further financing.[9]

---

8 Nasto, B. Chasing biotech across Europe. *Nature Biotechnology*, 2008. 26:283-288.

9 Shaw, G. Companies leave Canada because of tax laws. *Vancouver Sun*, December 7, 2007.

V

Conclusion

Once you understand the various legal, regulatory, political, commercial, and scientific factors that define, enable, and constrain the biotechnology industry, it is possible to apply this knowledge in many ways. This conclusion focuses on three common extensions of the previous chapters: building biotechnology companies, investing in biotechnology, and career development. Without repeating the content in the earlier chapters, these chapters integrate cross-cutting themes and provide further discussion on high-level elements.

Chapter 17
Building Biotechnology

Doriot's Rules of Investing

Seek companies demonstrating:
* New technology, new marketing concepts, and new application possibilities
* A significant participation by investors in management
* Staff of outstanding competence and integrity
* Products or processes that have been prototyped and have intellectual property protection
* Promise to enable an initial public offering or sale of company within a few years
* Opportunity for a venture capitalist to add value beyond dollars invested

General Georges Doriot of American Research & Development, the first dedicated venture capital firm

Successful biotechnology business development relies on three elements: technology, management, and capital. The basic value proposition of biotechnology companies is to develop applications based on proprietary technologies, granting them monopolies on the products of their R&D investments. Research and development are supported by management and capital. Management is responsible for identifying commercial possibilities for the markets a company wishes to target, and for positioning the company to realize them; capital is required to fund research and development and ultimately enable commercialization.

An examination of the qualities sought by offices experienced in launching biotechnology companies provides practical examples of factors important for long-term success. According to MIT's Technology Licensing Office, "[p]ositive indicators include very early-stage research, a technology that has several potential applications, no existing companies dominating the field, and an inventor who wants to participate actively in his or her invention's commercialization." They observe that emerging technologies with multiple new markets are often best exploited by focused and dedicated entrepreneurs who are funded by venture capitalists with an understanding of technical and business risk and reward.

Manchester Innovation Ltd., a technology transfer arm and incubator attached to the University of Manchester in England, requires that prospective clients complete the following objectives:

- Define the company's patent strategy and likely market potential for its products
- Write a robust business proposal
- Obtain at least seed finance
- Set up a board of directors and core team

They identify the most common blind spot for founders as a failure to realistically evaluate the market value of future products and acknowledge competitors.

Fundamental questions that should be asked in evaluating a venture are whether a sufficiently large market exists and if it can be profitably served. Unsubstantiated estimates such as "we expect to serve x percent of the $\$y$ billion market" indicate a fundamental lack of understanding of business development needs. It is not sufficient to simply assume acquisition of a proportion of an existing market. As illustrated in Figure 13.2 in Chapter 13, an in-depth analysis is required to measure the size of the reachable market. A bottom-up analysis is also essential to account for the key actions and costs involved in acquiring market-share and serving a market.

Many entrepreneurs also fail to account for all existing and potential competitors, or their responses to new market entry. Some competitors may already serve the target market, but potential customers may also be using alternative solutions. These alternatives may not be immediately apparent. Even products addressing unmet needs may face competition; horse-drawn wagons were competitors to the first cars. Additionally, products for previously untractable or unidentified problems may face challenges in convincing potential customers that the unmet need exists and the solution being offered will work.

Potential competitors, represented by future technologies, must also be considered. Even with a complete inventory of competitors, a common mistake is to assume that competitors will act rationally.

BUSINESS MODEL

Biotechnology business models (described in greater detail in Chapter 10) can be segmented into a few discrete types, each with characteristics that suit them for specific contexts. Factors such as technical challenges, barriers to entry, and the level of competition in a sector (see *Porter's Five Forces* in Chapter 13) may dictate the best model for a new entrant or incumbent firm.

In an environment where barriers to entry are relatively low—no "gatekeepers" controlling markets through broad patents or domination of marketing channels—and the financing climate is amenable, dedicated product development may be a good strategy. As markets mature, control of markets through patents and ownership of key infrastructure or sales and distribution channels limits the ability of new entrants to reach customers, making tool or platform approaches preferable. Other situations, such as a glut of viable drug leads favor "no research, development only" (NRDO) models (see *Specialty Pharmaceutical / NRDO Models* in Chapter 10).

There are often several modes by which an invention can be commercialized. The processes of selecting a business model and attracting funding are linked. Some of these may be more lucrative than others, but attracting funding may ultimately require crafting an opportunity that suits market trends and the interests of investors.

When technological uncertainty is high, or when funding is particularly challenging to obtain, hybrid approaches may be favorable. A company developing a new drug screening method, for example, may prefer to initially focus on refining that technology and licensing it to industry partners. These licensees can offset R&D expenses

Table 17.1 *Biotechnology business models*

General models	Characteristics
Product development	High risk, high reward. Requires supportive financing environment.
No research, development only (NRDO)	Reduced risk, high reward. Dependent upon ability to acquire drug leads; lack of internal R&D challenges decision-making and long-term growth.
Reagents and tools	Low risk, low reward. Subject to commoditization and obsolescence; success is predicated on dominating markets and niches.
Service provider	May deliver value by aggregating technologies from multiple companies, or offer economies-of-scale by ensuring full-utilization of expensive equipment.
Special models	Characteristics
Hybrid product / platform	Reduces risk of product development, allowing a company to prove technologies and generate revenues. Potentially distracting to management and R&D efforts. May be used to distract investors to failing core activities.
Virtual company	Effective for bootstrapping start-ups, difficult and expensive to manage at later stages.
Non-profit	Requires charitable donations and ability to license viable abandoned leads from incumbents.
Repurposing	Similar to NRDO. Requires ability to license and patent existing drugs for new uses.

See Chapter 15 for more detail on business models

while vetting the technology. Once the technology has been proven by external partners, the licensor will be able to cite these cases to financiers, reducing risk and facilitating funding.

FIRST STEPS

With an understanding of the fundamentals of biotechnology business development and a commercial idea, the biotechnology entrepreneur is faced with the challenge of how to proceed. What should one do first: Write a business plan? File patents? Assemble a management team? Raise money? Perform critical proof-of-principle research? Completing these critical steps in an appropriate order is essential for success. Bad timing can lead to false starts or loss of commercial opportunity.

Technology is first and foremost in biotechnology. Biotechnology is innovative by nature, so the goal of most biotechnology companies is to produce products and services that satisfy market needs and generate profits. But technology is expensive to initially develop, and tends to be less expensive to copy, so it is necessary to protect new technologies through patents, trade secrets, employee confidentiality agreements, or other means.

Because a competitive advantage is central to commercializing biotechnology, this must be secured first. Without possession of a competitive advantage, it will be difficult to attract talent and funding. It is important to validate a competitive advantage. What may

seem at first to be a patentable invention may have been previously published or patented by another party. Likewise, an invention developed during or even outside of working hours at a previous employer may still be property of that employer.

While patents are commonly used to secure a competitive advantage, there are other possibilities. Trade secrets are an alternative means to protect inventions; invention assignment, non-disclosure, non-competition, and non-solicitation clauses in employee and consulting contracts can prevent leakage of this information. A company can also obtain an exclusive or limited license for an unexploited technology, a common method in

Figure 17.1 *Building a biotechnology company*

university and corporate spin-offs.

With a protected idea, the next objective is to assemble a management team. Investors, seeking a return on their investment, will demand evidence that a company can succeed. While technological abilities and market opportunities can predict how successful a company may be, skilled management is essential to realizing commercial goals. Therefore, investors will either demand evidence of capable management, or will use their own resources to locate and install necessary talent.

Because of the central importance of funding, a balance must be maintained between supporting long-term commercial goals and meeting the relatively shorter-term needs of investors. The primary causes of biotechnology company failure are mismanagement and undercapitalization. Accordingly, biotechnology companies should secure able investors and ensure that performance and economic-based milestones are consistently met.

SELECTING OPPORTUNITIES AND BUSINESS PLANNING

> It is not the planning that is important; it is the planning that makes you able to change it.
> *Dwight D. Eisenhower*

To attract investors and build a biotechnology company, a commercial idea must exist with the potential to generate revenues. Innovative technologies and ambitious goals may attract press attention and early investors, but continued success requires profitable commercial execution.

Consider the case of Genentech. The company quickly completed proof-of-principle research to demonstrate the power of their revolutionary bacterial protein-expression system and signed a development contract with the insulin market leader to produce human insulin. After developing a method to produce human insulin in bacteria, Genentech applied the same basic techniques and secured external funding under more favorable terms to produce human growth hormone. While Genentech has since diversified into additional research areas, their initial plan focused on a single, patent-protected, new technology with multiple defined commercial possibilities (see Box *Genentech: Commercializing a new technology* in Chapter 2).

The primary objective of a biotechnology start-up is to secure a competitive advantage, followed by assembling a management team and obtaining funding. At some stage in this process, a business plan will have to be formulated.

Business plans often change as a company develops. One of the objectives of formulating a business plan is to determine the best course for implementation. While the initial business plan may be a rudimentary outline of a commercial idea, investigations into the feasibility, applications, and market potential of the idea can lead to modifications, resulting in a relatively stable concept that justifies the formation of a company. If an idea has a small market, requires an extremely large investment, or can only be used as part of a third party's patented process, it may be better to sell or license the idea to appropriately positioned firms.

Commercializing biotechnology is a long and challenging process. It is essential to formulate at least a rough plan for commercialization before committing too many resources. In evaluating a core technology, look for intellectual property protection and the ability to generate multiple products or services. Applications should be judged by the criteria presented in Chapter 12: freedom to operate, availability of technological factors, and ability to generate a profit. Before investing too many resources in research and development, it is vital to develop an understanding of who the customers are and how they can be reached. In developing a company, look to Doriot's rules presented at the beginning of this chapter and the criteria sought by university incubators. These are the elements that investors will look for. Why wait until you meet investors to justify whether an idea is marketable?

REDUCING RISK AND MAKING MONEY

The risk-tolerance of investors and their desire for either low-risk or high-return investments (these are not necessarily mutually exclusive) change with market sentiments and ultimately influence

Box
Want biotechnology funding? Use a shotgun!

One of the challenges in attracting funding is finding an investor aligned with your company's future directions. Venture capitalists and angels may be interested only in large companies, in small companies, in diagnostics, in drug development, etc. One of the challenges, and opportunities, for small firms, is that their future is very uncertain. This makes it difficult for founders to decide how to pitch their company to potential investors, but it also makes it possible to pitch more than one version of the company. Not all these versions may align with the founder's visions, but if they facilitate funding they may be a worthwhile digression.

Rather than limiting the set of potential investors to those whose investment criteria are compatible with the founders' visions, it may be preferable to leverage a company's core elements to appeal to a wider variety of investors. Using a shotgun approach, it is possible to craft several business plans based on a common set of resources, each aimed at a different kind of investor, and to simultaneously pitch these different plans to appropriate VCs.

For example, a firm developing drugs based on a proprietary technology can be pitched as a drug development firm, a platform-licensing firm, or both. It can even be split into two entities—one focused on licensing the platform technology and the other as a licensor of the platform for specific applications. A firm focusing on drug development will have a very different structure than one focusing on licensing. The former will need extensive partnerships and resources to complete development, whereas the latter can be far leaner and focus on using partners to vet the technology to promote further licensing or to facilitate future funding for drug development.[1] Therefore, even if the founder's desire is to focus on drug development, a near-term platform focus can attract funding for later drug development.

The key element is to be creative in seeking funding and avoid unnecessary elimination of options. Most investors will realize that there is a great deal of flexibility in a start-up's business model and business plan, because there are so many unknown elements in a young company, and pitching to more investors increases the likelihood of attracting an investment.

---

1 A similar strategy has been used successfully by Domain Associates' Eckard Weber: http://invivoblog.blogspot.com/2007/08/one-two-punch-in-venture-capital.html

how much equity an investor will seek in exchange for a given investment: the cost of capital.

Risk is inexorably tied to revenues and profits. Investors in biotechnology companies seek assurances that they will receive a return on their investment. Investors will seek greater portions of equity commensurate with the perceived level of risk, which influences their expected return on investment (see *Valuation* in Chapter 11).

In risk-averse markets an effective way to reduce risk is to demonstrate the ability to generate revenues, and preferably net income, as soon as possible. This may require favoring near-term outcomes at the expense of long-term objectives by reducing R&D expenditures, seeking approval for smaller and/or safer markets, or selling lead compounds. In risk-tolerant markets, where investors may expect greater returns, a focus on short-term revenues can jeopardize long-term profitability by distracting management and R&D efforts from long-term value creation activities. The best way to attract financing in risk-tolerant markets may be to eschew organic growth in favor of large markets, synergistic acquisitions, and bold goals, in an effort to meet the high rate of return expected by investors.

A desire to avoid equity dilution motivates the pursuit of equity-free financing options (described in the section *Other Funding Sources* in Chapter 11), although in many cases equity financing can provide substantial benefits in terms of speed and flexibility. Founders may also seek to reduce the size of investments they are looking for in an attempt to control dilution. This strategy may backfire, as it

Table 17.2 *Raising funds to reach developmental milestones*

Milestone	Funding stage	Funding amount / burn rate
Proof-of-principle	Seed	$300-600k 3-6 months burn
Prototype	First-round	$1-5 million 1 year burn
Early product development	Early mezzanine	$5-15 million 1-1.5 years burn
Commercial launch	Late mezzanine / IPO	$20-50 million 2-3 years burn

Source: Birndorf, H.C. Rational financing. *Nature Biotechnology*, 1999. 17:BE33-BE34

can leave a company with insufficient cash to reach the next funding stage. In a famous example, Hybritech, one of the first biotechnology companies, asked for \$178,000 in seed funding to fund research in preparation for first-round funding. The investors, sensing that more money was needed, gave Hybritech \$300,000. Asking for too little money can also discourage investors. As described in the Box *Venture capital: The poker analogy* in Chapter 11, investors are seldom shy about funding strong companies and often seek to invest as much as possible at the most attractive entry points as a means to maximize their returns. Asking for too little money may indicate modest goals (and, by extension, modest outcomes), encouraging potential investors to seek other opportunities.

MAXIMIZING MULTIPLES

The ultimate goal of biotechnology companies is to use molecular biology techniques to produce novel products worthy of a price greater than the investment in developing those products. While it was once possible for nascent biotechnology companies (e.g. Amgen and Genentech) to develop into fully integrated research, development, and commercialization enterprises, market dynamics have made it much more cost-effective for companies to specialize in discrete elements of this pathway. To be profitable, it is essential for R&D firms to focus on maximizing multiples: developing products and selling them when the cost of further development exceeds the value created by additional R&D, and when further development can be more efficiently and effectively performed by another entity.

In developing a biotechnology product it is vitally important to be cognizant of the actions necessary to reduce risk and increase the value of products in development, and the cost of those actions. Cost, in this case, is not directly measured in dollars, but rather by time and loss of equity.

While equity-free funding is available for biotechnology development, it tends to be restricted to early-stage activities. Development-stage activities generally require exchange of company equity for funding (see *Development Stages and Funding* in Chapter 11). Figure 17.2 shows a general scheme of biotechnology value creation

exit zone

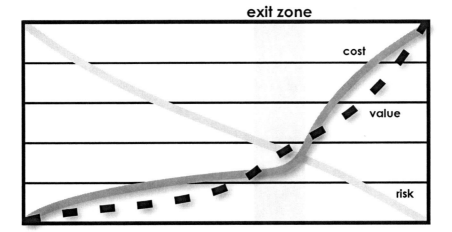

concept > patent > pre-clinical > clinical trials > approval

Figure 17.2 *Value creation in biotechnology*

in concert with risk reduction and cost for a nascent biotechnology firm, in a hypothetical environment where the returns on investment are optimal for exit in early clinical trials.

As a commercial idea matures from concept to tangible product, the risk of the opportunity decreases. The development cost naturally increases during development, as does the value of the product in development. The factors determining cost, value, and risk are independent of each other (a decrease in risk does not necessarily produce a commensurate increase in value), but are dependent upon the product being developed, a company's business model and resource base, and market conditions determining cost of capital and the value of products in development.

In this very simple example, the cost of proceeding beyond early clinical trials exceeds the value creation of these investments, making early-clinical trials the optimal point to sell a product (e.g., selling Phase I or Phase II leads). Selling prior to the "exit zone" (prior to early clinical trials in this example) is an unfavorable option, because the value of the hypothetical products is less than the cost incurred to proceed to that point of development—not enough risk has been removed from the investment. Selling after the exit zone (after early clinical trials in this example) is also unfavorable, because the costs of proceeding past the exit zone exceed the increase in value—while the risk has been sufficiently reduced, the capital expenditures are too great. The company in this example would therefore be best served by licensing or selling drug leads that have passed early clinical trials to parties who can perform late-stage development at lower costs.

INTEGRATE MARKETING EARLY

A common cause of product failure is poor integration of marketing with R&D decisions. Biotechnology is innovative by nature, so many products get developed based on availability of new technologies. This is termed a "technology push," and can lead to development of products with no market. So it is important to start from the endpoint—sales—and consider market needs and demands in directing product development. Many investors and entrepreneurs fail to consider marketing and reimbursement issues, focusing on actions required to sell drug leads or entire companies well in advance of commercialization. This myopic view can lead to problems when the time to sell comes, as downstream partners or investors may be unwilling to consider technologies with poor market potential.

Some of the questions which should be asked early in development are:

- Is there a market willing and able to pay for potential products? What is the path to the market?
  - See Box *Biotechnology myth: Build it and they will come* in Chapter 13
- What regulatory incentives can you leverage?
  - See Box *Genzyme: Building an enterprise on orphans* in Chapter 8
- What resources and partnerships will be required to complete development and reach the target market?
  - See Box *Exubera: When your partner doesn't sell* in Chapter 14

In some cases it is not possible or relevant to answer all these questions at the outset. For novel products, certain marketing issues may not be known until the product is launched. For early-stage products or those serving dynamic markets, it may not be appropriate to dedicate too much time to measuring markets, as the target markets and products themselves may change over the course of development. However, in these cases it is still important to track marketing issues and to continually reassess them as products mature. Failure to do so may result in development of products for markets that don't exist, or for markets that don't need them.

EXITS AND OPTIONS

It is vital, at a very early point in a company's development, to formulate a path to maturity or exit, and to start planning for unforeseen obstacles. Just as excessive focus on developing technology, to the exclusion of marketing, can lead to failure, it is also important to consider the path a company will take. Some of the questions which should be asked while planning for growth, exits, and contingencies are:

- What if your lead project fails?
  - See Box *Dangers of not having a pipeline* in Chapter 12
- What if your product doesn't have the properties you hoped it would?
  - See Box *Flavr Savr tomatoes: Operating in unfamiliar markets* in Chapter 15
- What if the demand for your product exceeds your capacity?
  - See Box *Enbrel: Underestimating market demand* in Chapter 13
- What if you lack freedom to operate?
  - See Box *Amgen v. Transkaryotic Therapies: Strategic patenting* in Chapter 7
- How do you resolve a valuation disagreement
  - See Box *Remicade: Resolving valuation disagreements* in Chapter 14

- How would you prevent or manage a product recall?
  - See the section *Off-Label Use* in Chapter 8 and Box *Preventing a product recall from bankrupting a company* in Chapter 15
- Do you have a plan for succession?
  - See Box *Replacing founders* in Chapter 10
- What if the IPO window closes?
  - See Box *Alternative route to going public* in Chapter 11

These plans do not need to be rigid or exclude other options, but they are essential to help set future goals and measures of progress. A company seeking to be acquired, for example, should develop in a different manner than one seeking an IPO. The former should focus on developing products which could demonstrate near-term value to acquirers, while the latter should place more emphasis on demonstrating the capacity for long-term growth. Failure to plan for growth and commercialization (or exit) can result in disorientation and lead to failure. Operating without a set of goals and plans to achieve those goals can leave a company unable to make important decisions, and may also motivate investors, partners, and employees to seek better-defined opportunities.

Chapter 18
Investing

If you don't know jewelry, know the jeweler.
Warren Buffet

CAVEAT EMPTOR

Less than an hour after shares for Genentech's initial public offering (IPO) opened at $35 in 1980, the price had appreciated to $88, making for one of the largest stock run-ups ever, and casting biotechnology in investor's minds for decades to come.

Investors are still drawn to biotechnology today because of the high profit margins, years of patent-protected sales, and the ability to address pressing needs in growing markets. However, a measure of perspective must be applied in evaluating the potentials of investing in biotechnology.

Most biotechnology companies make for poor investments. Relatively few of the multitude of biotechnology companies eagerly developing innovative products to address lucrative markets will ultimately succeed. Of those few that do succeed, only some will be profitable, and only some of those will deliver outstanding returns. What makes investing in biotechnology so compelling is the historical ability of just a few companies to deliver such outstanding returns that they effectively prop up the entire sector.

A ten year follow-up analysis of the 41 biotechnology companies that went public in 1995 and 1996 found that:

- 25 were still independent

- 4 had gone out of business
- 12 had been acquired
- 7 had been acquired at a loss
- 6 were profitable
- 15 had a share price higher than the close on their first day of trading[1]

This basic analysis highlights an important trend. Even among the select biotechnology companies that are able to mature into public entities, most fail to deliver positive returns over an extended period of time. According to Eaton Vance Worldwide Health Sciences Fund manager Samuel Isaly, only 40 of the 1,000 biotechnology companies that have gone public in the history of the industry have ever attained profitability.[2]

INVESTING IN BIOTECHNOLOGY

Despite the numerous challenges of investing in biotechnology companies, investors have the ability to participate in the development of products that benefit humanity and may also realize outstanding financial returns. To invest profitably in biotechnology companies, one must appreciate the influences of scientific, legal, regulatory, political, and commercial factors. While there are a variety of strategies for investing in public markets, in developing a biotechnology investment strategy it is vital to consider the unique characteristics of biotechnology companies and the challenges they face.

Biotechnology companies are very research and development-intensive. Start-up companies are especially risky investments because they have high burn rates and sell on the promise of future profits. For mature and start-up companies alike, the success of any individual project is far from certain. Research projects can fail at any stage for predictable or unpredictable reasons. The great uncertainty of whether or not product development will be successful challenges

1 Travers, C. Grading old-school biotech. *Motley Fool*, January 25, 2005. http://www. fool.com/news/commentary/2005/commentary05012506.htm

2 Jacobs, T. Great company, bad stock. *Nature Biotechnology*, 2005. 23(2):173.

the formulation of financial projections, limiting the predictive ability of traditional valuation methods.

Faced with the inherent uncertainty of biotechnology development, investors must learn to accept the risks associated with investing in biotechnology companies. This book has described the scientific, legal, regulatory, political, and commercial factors specific to biotechnology companies. Understanding the importance of all of these subjects in biotechnology company development allows the educated investor to objectively assess an investment opportunity.

While identifying successful companies enables investors to profit from breakthroughs in biotechnology, poor stock selection is unlikely to yield any gains. An investor who does not appreciate the unique challenges and opportunities of biotechnology research is better off purchasing general sector mutual funds, investing in the companies that stand to indirectly benefit from biotechnology, or avoiding biotechnology investments altogether.

STRATEGY

A fundamental difference between investing in biotechnology companies and investing in more traditional companies is that traditional analysis methods are less able to predict success in biotechnology. This is particularly relevant for companies without products on the market. Biotechnology product development is fraught with unexpected failures. Projects may fail for any number of reasons that could not be predicted at the outset. Share prices, especially those of smaller companies, may fluctuate significantly in response to project development progress. Significant R&D successes can see stock prices appreciate rapidly; disappointing results can lead to precipitous price drops.

It is necessary to consider multiple factors to minimize investment risk and predict the future prospects of biotechnology companies. Highly-focused analysis of revenues and expenditures, for example, is of limited value for an early-stage biotechnology company with characteristic high R&D investments and the future potential of substantial income. It is important to assess the probability of achieving a future revenue stream in addition to measuring the possible magnitude of future revenues.

FUNDAMENTALS ANALYSIS

A general survey of desirable characteristics in biotechnology companies reveals the following criteria for promising investments:

Corporate Qualities
- Maturity
- Business model
- Experienced management
- Competitive advantage
- Institutional support
- Favorable financial analysis
- Sufficient funding to fund operations for several years

Product Qualities
- Successful products on the market
- Numerous products in development
- Products that target large or underserved markets

The above criteria are subjective, requiring a fair amount of personal judgment in evaluating the quality of a company's strengths in each area. Furthermore, the above criteria do not necessarily lead to success but can identify likely failures. A company with excellent fundamentals may fail, but a company with poor fundamentals cannot be successful. It is therefore important to not obsess over individual measures, using the criteria instead as a framework upon which to structure an overall assessment.

CORPORATE QUALITIES

MATURITY
The maturity of a biotechnology company is an important consideration in assessing corporate strength and stability. Summarizing the grouping of biotechnology companies by maturity and sta-

bility as presented in Chapter 10, biotechnology companies can be roughly grouped into three categories. Established large-cap firms with positive revenue streams are labeled mature. Those with strong fundamentals and excellent prospects for near-term profitability are labeled promising, and the remaining biotechnology companies, those without near-term certainty of sustainable profits, are labeled emergent.

Briefly, mature companies are the most amenable to traditional financial analysis. In addition to a mature company's historical record and current financial position, it is also important to consider non-financial factors such as the business model, management track record, quality and quantity of partnerships, and apparent strength of research efforts in order to project future performance. Mature companies usually have a number of successful products on the market as well as proven ability to maximize revenues and protect market share.

Promising companies are more difficult to objectively evaluate than mature companies. The future prospects of these firms often depend on events such as success of key clinical trials, the outcomes of which cannot be reasonably predicted. Evaluating revenue streams of promising companies is less predictive of success than it is for mature firms, because the expectation is that the revenues of promising companies will grow significantly in the future. A more important factor in assessing promising companies is financial health. In order to succeed, a promising company must have sufficient finances, or the potential to raise sufficient finances, to fund operations and growth in order to develop into a mature firm. Other important considerations are the quality of non-financial parameters such as the business model, management expertise, quality and quantity of partnerships, and strength of research efforts. While it is difficult to predict revenues for sales of yet-to-be developed novel products, it may be productive to project revenues and use these figures in a traditional financial analysis to determine if a company's current stock price already reflects positive future expectations.

While investing in promising companies involves significant risk, emergent companies present even more risk. Financial analysis of an emergent firm using projected revenues may be an effective way to determine if a company's value exceeds even optimistic revenue projections, but the uncertainty of revenues for these firms limits the utility of this exercise. Furthermore, emergent company business plans, business models, and target markets may change significantly prior to development of a stable revenue stream. The unpredictability of future research directions and revenue possibilities means that the most effective way to invest in emergent companies is to accept a large component of risk and look at fundamentals that provide a strong research infrastructure such as access to capital, quality of the management team, quality of partnerships, and aggressiveness of research efforts.

BUSINESS MODEL

One of former Fidelity Magellan Fund manager Peter Lynch's investing principles is, "never invest in any idea you can't illustrate with a crayon." It is vitally important to understand, even at a superficial level, how a company is going to make money. It is easy to be enticed by innovative technologies or large markets, but one must rationalize how an investment can increase in value. Promising companies may employ innovative technologies or seek to serve lucrative markets, but innovative technologies and products do not create revenues; sales do.

MANAGEMENT TEAM

The quality of a company's management team is arguably the best predictor of success. This is especially true for young companies. Individual research projects may succeed or fail, but it is ultimately up to management to secure resources to enable research, to guide research towards profitable ends, and to facilitate the profitable sale of developed products and services. Managers with a history of success are likely to succeed again.

A company's managers should have experience in general skills such as managing collaborations and securing financing, as well as in areas pertinent to the specific development plan such as unique regulatory requirements or development hurdles. Past experience dealing with the issues that a company is likely to face prepares management to predict and swiftly resolve problems.

Some history of failure among managers is not necessarily an indication of unsuitability. Ironically, failure at a previous firm can be a more informative experience than success. A share of failure keeps people humble. Failures also teach people to do things differently and can provide insight into which actions should and should not be taken in unfamiliar situations. One of the factors for the successes of American entrepreneurs in biotechnology and other domains is attributed to the relative willingness among investors to back people who have failed in previous ventures.

COMPETITIVE ADVANTAGE

As described in Chapter 7, biotechnology companies require a competitive advantage to prevent competitors from capitalizing on the efforts of pioneers and denying them the ability to recover investments in research and development.

The number of competitors and the developmental maturity of competing products are important considerations in assessing the value of a company's products. The safety, efficacy, and price of competing products are also important factors in predicting the impact of competition on market share.

For biotechnology companies, patents are the most commonly used means to secure a competitive advantage. Patents grant innovators 20 years of exclusive rights to exclude others from making, using, offering for sale, or selling an invention. Patenting an invention requires the innovator to disclose the best means to practice an invention, which can facilitate the emergence of competitors upon patent expiration or invalidation. Key technologies should be protected by multiple patents to protect against invalidation of a single patent and to cover as broad an application area as possible. Beyond simply restricting discrete applications, an additional benefit of employing multiple patents is that they can also convey a willingness to aggressively defend intellectual property. In one example, Affymetrix used more than 400 patents to protect its microarray technology, raising a significant legal and scientific barrier to competitors. The great cost of legal expenses to simply determine freedom to operate in the microarray space likely dissuaded many potential competitors.

A law degree or Ph.D. is not necessary for a general assessment of a company's patent strength; examining a company's history can be telling. A company with a history of winning patent decisions is likely to continue to succeed in protecting its intellectual property. Furthermore, an established reputation for securing favorable judgements can discourage infringers and encourage favorable out-of-court settlements.

While patents are not the only way to secure a competitive advantage in biotechnology, they are the most common. Additional forms of competitive advantage include trade secrets, restricted access to superior sales forces, distribution networks, and lucrative partnerships. In the final analysis, it is important to assess the ability of any competitive advantage to exclude competitors and protect profits.

INSTITUTIONAL SUPPORT

Partnerships provide both cash and endorsement, and are of great importance in the biotechnology industry. Partnerships with industry leaders indicate that knowledgeable and capable industry insiders endorse a junior partner's scientific and commercial possibilities. Because large companies and venture capitalists often respectively partner with, and fund, multiple companies with the knowledge that they can cut poor performers loose, it is important to consider institutional support in the context of other company fundamentals.

In assessing the quality of a partnership one should look for cash and rights distributions. The magnitude of upfront payments, milestone payments, downstream royalties, and co-promotion rights are all measures of the value of a partnership. Large cash commitments, especially upfront payments, indicate strong support. Furthermore, the obligations of each partner (e.g., does the junior partner have to simply identify leads, or must they produce a drug that passes clinical trials?) indicate the relative competitive strengths of each partner. The relative stake of each partner dictates how any profits will be distributed.

Just as partnerships with esteemed firms represent an assessment of a company's value, the quality of the money behind a start-up is another measure of industry insiders' assessment of a start-up's prospects. Backing from a top-flight venture capital firm with experience in biotechnology is an excellent indication that some very knowledgeable people think that a company's prospects are good. One way to assess the quality of a financial backer is to look at their track record. A backer with a history of successful investments in biotechnology is likely to experience future success. Backing from major corporations with relevant industry experience is another positive indicator.

FINANCIAL ANALYSIS

While the other biotechnology company evaluation criteria in this section describe methods to assess whether or not a company is likely to succeed in product development, financial analysis is also important because it provides information on whether a stock's value represents the value of a company. A company may have strong prospects, but it is also important to determine if the stock's value represents, or even exceeds, expectations. Because many companies lack significant revenues, financial assessment of biotechnology companies is challenging. Even those companies with sizable revenues often have uncertain futures.

An overview of valuation methods is presented in Chapter 11. A number of other valuation models have been proposed over the span of decades to determine the value of biotechnology companies. Models for success in biotechnology have considered factors such as the number of Ph.D.s employed, market forecasts, cash flows, and the ratio of R&D expenditures to earnings. Unfortunately, a method to effectively reduce biotechnology development and market potential to mathematical formulas does not yet exist. Selecting peer groups, projecting revenues, and assigning development risks are subjective measures. Calculations based on these figures are likewise subjective. Financial analysis can provide a useful metric for biotechnology, but it requires careful integration with non-financial parameters to increase relevance.

Peer comparison is a method used to determine if a given company's share price is relatively more or less expensive than other similar companies. Peer comparison can also be used to examine whether certain biotechnology sectors are relatively over- or under-valued relative to other sectors. The simplest tool for peer comparison is calculation of the ratio of stock price to annual earnings. This so-called price to earnings (P/E) ratio is calculated by dividing a company's current share price by the earnings per share for the previous twelve months. P/E ratio calculations are of little relevance for companies without earnings.

There are a number of other peer comparison measures which are more sophisticated than P/E ratios, but they all share a common weakness. The assembly of a peer group requires a certain amount of subjectivity. Furthermore, biotechnology peers rarely develop similar products for similar markets, meaning that peer differences are expected and will be based on non-financial attributes. Most importantly, peer comparison only yields relative valuations. A company may be undervalued or overvalued relative to its peers, but if it is in a misvalued sector its share price can nonetheless fall or rise in response to market sentiments rather than its intrinsic value.

BURN RATE

For many biotechnology companies, product launch and profits are often many years away and require significant financial investment. Access to cash is an important factor in enabling a company to complete development, making a company's rate of spending, or burn rate, an important measure. Stock analysts often look for companies with a minimum of two years' cash reserves. This time span should either permit the completion of milestones that facilitate future funding, or lead to profitable revenue streams.

PRODUCT QUALITIES

PRODUCTS IN DEVELOPMENT / ON THE MARKET

Products on market are not only a source of revenue, they are also a testament to management's ability to guide a product through development and commercialization. Companies with product revenues can use this incoming cash flow (see Box *Focus on free cash flow* in Chapter 11) to maintain operations and to fund future development. The ability to draw upon internal resources also means that these companies will likely be able to maintain greater independence in future product development and retain a greater share of profits.

For companies without successful products, those with products in late stages of development present less risk than those with products in early stages. The process of clearing a potential drug for clinical trials and proceeding through the three phases of clinical trials to demonstrate safety and efficacy is very challenging and unpredictable. The more advanced a company's products are in this process, the lower the risk of significant setbacks or failure (see *Maximizing Multiples,* and *Reducing Risk and Making Money* in Chapter 17).

In judging development progress, it is important to consider the source of a report. Many journalists struggle to understand the fundamentals of biotechnology research and may inadvertently make errors in reporting. One should examine whether a report comes from a respected newspaper, magazine, or broadcast, and if it is produced by individuals with proven expertise. Furthermore, while the FDA regulates press releases about drugs that have been approved, there is relatively little oversight of press releases about experimental drugs. Reports based on single Phase I trials, for example, may project enormous market potential well before the safety and effectiveness of a drug is established.

PIPELINE DIVERSIFICATION

Biotechnology product development is fraught with unexpected hurdles, setbacks, and failures. Accordingly, it is imperative that companies have multiple products in development to provide alternatives should individual research lines fail. Multiple products are essential to support long-term growth.

A company's development pipeline should ideally be sufficiently broad to address markets with good revenue potentials and to provide a measure of stock price stability. Excessive diversification can be as detrimental as a lack of diversification. A lack of focus in research projects demonstrates poor management. Furthermore, a company pursuing too many unrelated objectives may be unable to focus sufficient resources to overcome significant development complications or enable commercial success (see the section *Diversification and Focus* in Chapter 12).

LARGE OR UNDERSERVED MARKETS

Significant long-term value can be derived from products which are sold to a broad customer base. This is the reason why drugs, food products, and health services are common applications for biotechnology development. Another factor in assessing the quality of a product is evaluating how pressing a need it serves and how frequently it is likely to be purchased. For example, a drug serving a chronic life-threatening condition is likely to generate greater revenues than an infrequently used or non-essential drug.

Products in development should have defined markets and clear advantages over any existing alternatives. It is important to not overestimate the potential of poorly defined products serving defined markets or defined products serving poorly defined markets. Celera faced a crisis after completing its goal of producing a rough draft of the human genome when it found that the market for genomic information was insufficient to support desired growth rates. This misjudgment required Celera to divest itself of its genomics business to focus on drug development, before shifting focus again to creating molecular diagnostic tests.

It is also important to assess the barriers to successful development:

- How well is a problem defined?
- How refined are the tools needed to solve the problem?
- How much research and development is necessary?
- If a company is developing applications of a promising technology that has eluded others for years, then what are the odds of success? What new tools or techniques are they using?
- Has a recent scientific or technological development created new opportunities? Is there a way to vet the potential of this recent advance?

A poorly defined problem introduces the possibility of large, unpredictable, capital requirements (see Box *Poorly defined diseases discourage drug development* in Chapter 4). The availability of appropriate tools to develop an application reduces the uncertainty of capital and time requirements and allows for defined and predictable milestones. It can be difficult to determine which new technologies are best able to reduce uncertainty, but one positive metric is the entry of new players using a new technology. Applications with great potential will likely attract multiple developers. Revolutionary technologies such as gene splicing, monoclonal antibodies, gene therapy, and RNA interference all led to the emergence of new companies focused on leveraging them. Not all of these technologies delivered on their initial promise, but weak adoption can be a good sign that a technology is unlikely to deliver great returns.

Beyond the qualities of individual products, it is also important to consider a company's prospects for future development. For example, a company which has been able to consistently produce moderately valuable products may fare better in the long-term than a company which infrequently produces more valuable products.

INVESTING ON TRENDS

In addition to assessing the fundamentals of companies, it is important to consider market trends. The volatility of biotechnology stocks challenges the interpretation of trends and the application of this information, but there are some important factors to consider.

When market support is strong, undeserving companies often see their share prices rise as they are associated with market leaders. When market support weakens, even good companies will see their prices unfairly depressed. In the long run, companies with strong fundamentals are likely to succeed, whereas companies with poor fundamentals may be acquired, liquidated, or stagnate. This reality underscores the importance for long-term investors to evaluate fundamentals.

Short-term investors who purchase company shares primarily in the hopes of near-term FDA approval should consider the possibility of a share price collapse should a drug be found unapprovable. By investing early, investors may be able to secure relatively greater returns, but they also face significant risk. MSN Money examined the performance of companies following FDA review of sixty biotechnology drugs from 1998 through 2000. They found that 55 percent of the tracked stocks were trading higher thirty days after approval, with an average gain of 7 percent. Sixty days after approval the average gain was 13.4 percent, with 58 percent of the stocks trading higher. Ninety days post-approval 60 percent of stocks were trading higher; an average gain of 12.6 percent.[3] Conversely, shares can fall 70 percent or more in a single day following failure of a key drug to receive FDA approval. While this historical example is not indicative of future market performance, the potential to average a greater than 13 percent gain by buying drug stocks on the day of approval and holding them for two to three months represents a simple and potentially lucrative way to invest in biotechnology.

CONCLUSION

Any model to determine the value of a biotechnology company or likelihood of commercial success relies on a certain degree of personal opinion, informed or otherwise. It is important to be aware of the relative contributions of measurable and unmeasurable factors in assessing the merits of individual companies. Following meticulous financial analysis, for example, a company may be deemed undervalued or overvalued based solely upon which other companies it is compared against. Similarly, overestimating or underestimating the future potential of a product can effectively make all other measures insignificant. It is therefore necessary to consider whether the sum of measures forms a consensus; investment decisions should not be overly reliant upon individual measures.

Although assessment of the criteria described above cannot identify certain long-term winners, it can effectively identify losers. Consider a company that fails in multiple measurement criteria—poor cash position, an uninspiring management team, and weak technology—such a company is almost certain to fail. Even in the unlikely event that a lucrative breakthrough emerges, a poorly

---

3 Niederhoffer, V., Kenner, L. Biotech: an investing frontier for risk-takers. *MSN Money*, July 5, 2001.

equipped company will not able to capitalize on it.

Investors should take the time to learn about the unique aspects of biotechnology companies and reflect on them in making investment decisions. Investors who are unable or unwilling to make this commitment may fare better in selecting alternative methods to capitalize on growth in this sector. While small biotechnology companies may develop lucrative products, they are also likely to form research, production, or marketing alliances with industry leaders. Investing in the biotechnology and pharmaceutical leaders that share in the success of smaller firms is a relatively safer way to invest in biotechnology.

Another alternative is investment in sector mutual funds. These investment vehicles tend to roughly mirror the rise and fall of a whole sector and permit investment in biotechnology without the volatility associated with individual companies. Mutual funds range in investment focus from specific sub-categories of biotechnology such as genomics to broad funds covering healthcare in general. As with individual stocks, investors should only invest in mutual funds which cover areas that they understand. Investing in a niche fund without understanding the future potential of that niche is likely to offer the same poor returns as investing in a company without understanding its commercial prospects.

Chapter 19
Career Development

Choose a job you love and you will never work a day in your life.
Confucius

Although the scientific aspects of biotechnology may seem daunting to outsiders, biotechnology companies employ many of the same job functions as other firms. In addition to researchers and technicians, biotechnology companies also employ receptionists, lawyers, engineers, janitors, salespeople, accountants, and a host of other business professionals.

While a Ph.D. in science is not a necessity for a career in biotechnology, it is important to understand topics relevant to one's role and to be able to communicate effectively with others. A computer programmer on a bioinformatics project, for example, must know enough biology to communicate with biological scientists. Salespeople should understand the applications of their wares, permitting them to advise customers of useful products. While there are numerous opportunities outside of research and development for individuals without advanced degrees in science, the nature of R&D requires individuals with refined knowledge and skills.

EVALUATING POTENTIAL EMPLOYERS

The performance of a company and the sector it is in influences the job stability and role of its employees. For example, the genomics sector cut 1,500 jobs from January 2001 through June 2002 following loss of public market support.

Job candidates should spend at least as much time selecting an employer as one might spend selecting a stock for investment. Candidates should look at the quality and experience of a company's management team, technology of products being developed, cash position and soundness of financing, quality of investors, competitive advantage, market need, and competition. Asking about the tenure of the management team as well as future hiring plans can indicate the stability of a firm and suggest whether a hiring decision is based on growth or replacing lost functions. For start-ups, where answers to many of these questions are unavailable or unknown, discussions of a company's burn rate and plans for future financing can indicate job stability and the likelihood of being redirected to different projects. Asking about a start-up's exit strategy and time frame can likewise reveal information about potential job stability and impending transformative events.

Other important elements include benefits such as retirement savings plans and health insurance. Companies may also offer employees stock options to offset lower salaries or to encourage continued tenure. It is important to consider how and when these options can be liquidated. Stock options in private companies may only have value if and when the company becomes publicly traded, and in public companies the stock options may likewise have limited value if their purchase price exceeds the market price (due to the options being issued at a time when the market price was higher).

While companies are unlikely to divulge complete details to job candidates, asking the right questions can reveal corporate strengths and weaknesses, and can convey an understanding of the fundamentals of commercial biotechnology to interviewers.

JOB DESCRIPTIONS

RESEARCH ROLES

Individuals with advanced scientific knowledge and abilities conduct and guide research. The role and responsibility of researchers varies with their individual skills and expertise.

Laboratory technicians perform various maintenance tasks in

laboratories and may perform experiments as well. Most laboratory technicians have master's degrees, but opportunities are available for those with bachelor's degrees or high school diplomas. The functions performed by technicians range from maintaining stocks of reagents and research supplies to performing or supervising routine operations and performing supervised research experiments. Important skills for research technicians are communications and precise record keeping.

Opportunities for individuals with Ph.D.s, who may or may not have post-doctoral training, vary from research team members to core facility managers. Initial posts for young scientists are as members of research teams. After developing specialized expertise, seasoned scientists may lead research teams, plan and manage multiple research projects, and run labs.

The scientists who work in biotechnology companies generally have backgrounds in biology, chemistry, and medicine. A scientist's background influences which roles they will perform. Chemists, for example, will most likely find themselves engaged in early drug discovery. Developing potential drugs in preparation for clinical trials will likely involve pharmacologists, and running clinical trials requires physicians.

Non-research roles such as management positions, intellectual property management, and consulting require intimate knowledge of scientific fundamentals and research dynamics. Although individuals in these fields do not necessarily require experience in the specific field of research they are working in, advanced training such as a master's or Ph.D. degree in a related field is a great asset.

NON-RESEARCH ROLES

Aside from research and development, biotechnology companies perform many of the same operations as other companies. Directors and managers, for example, generally have business or management degrees (often in addition to scientific doctoral degrees). The central role of funding and financial management in biotechnology establishes a demand for individuals with proven financial expertise. Likewise, communications and human resources professionals are also needed for their specialized abilities.

Because of the importance of intellectual property protection in biotechnology, lawyers are needed to compose and prosecute patents and assist in the collection and evaluation of competitive intelligence. Marketing and sales experts are needed to study and develop markets and ultimately enable the delivery of products to consumers. Furthermore, the potential for substantial financial returns has attracted great interest for biotechnology in public markets, creating a demand for analysts, venture capitalists, and investment bankers with an understanding of biotechnology-related financial issues.

There is a strong need for specialists who can help with development and manufacturing processes. Engineers with skills in water purification, brewery design and operation, product packaging, and electrical and software design can all find roles in biotechnology production. Bioinformatics research likewise requires individuals with proven computer programming abilities who can apply their skills to biotechnology problems. There is also a strong need for branding and communications professionals. The significant negative impact of patent expirations on sales creates a great opportunity for individuals who can successfully develop strong brand positions for pioneers to help them sustain sales following patent expiration. The volatile nature of R&D and funding also necessitates carefully crafted communications with a diverse audience (see *Public Relations* in Chapter 15).

While a majority of individuals involved in research and development have advanced degrees in the sciences, there are numerous opportunities for individuals without scientific degrees. For example, the career path of Kevin Sharer took him from being chief engineer on a nuclear submarine to working at GE and MCI before becoming president and CEO of Amgen, a leader in the biotechnology industry. Prior to joining Amgen, his closest prior exposure to biotechnology was high school biology and college chemistry.

Regardless of educational background or expertise, the key to a career in biotechnology is to appreciate the unique challenges faced by biotechnology companies. By understanding the factors influencing biotechnology companies it is possible to select opportunities and position oneself for a rewarding career.

PH.D., MBA, OR BOTH?

A common question asked by students and business professionals seeking to enter the biotechnology industry is whether a Ph.D. or MBA would improve their career prospects. There are numerous opportunities for individuals with either, or neither, degree. The choice of what to study depends largely upon an individual's career interests and the desire and ability to complete advanced studies. The financial risks and rewards also vary by degree attainment and career path.

Research managers should generally have a Ph.D. in science or a medical degree. Other roles such as marketing, accounting, sales, and human resources do not require an advanced degree in science, but benefit from a business background with an understanding of pertinent biotechnology industry issues.

Because biotechnology companies focus on researching and developing products based on advanced scientific principles, Ph.D. degrees are common among researchers and managers alike. Pharmaceutical companies, with a relatively greater involvement in marketing and licensing, require managers with business expertise in addition to those with scientific backgrounds.

One way to assess the career potential of specific degrees is to examine job postings and the credentials of individuals in positions of interest. Many companies list the academic qualifications of their senior management. Job postings also describe necessary qualifications for specific positions. It is also instructive to look at the non-academic requirements in job postings and the career histories of executives.

Individuals interested in obtaining both a Ph.D. and MBA may find that the preferred route is to obtain a Ph.D. first. While both these advanced degrees require dedication and passion to successfully attain, completing a Ph.D. requires sustained dedication and working long hours for more years than MBA programs require. Beyond time spent in lab, students should be passionate and spend most of their waking hours thinking about their research. The need for dedicated focus and intense time demands are reasons why most first year doctoral students are relatively young. It is accordingly far less common to pursue a Ph.D. after an MBA than to follow Ph.D. studies with an MBA (another possible reason is opportunity-cost: salaries for fresh MBA graduates are generally higher than for fresh Ph.D. graduates).

It is also worth considering other educational options. An alternative to a Ph.D. is to pursue a bachelor's or master's degree. A certificate program may likewise be a reasonable alternative to an MBA. An increasing number of schools are also offering blended programs, combining science and management classes. In the final analysis, the range of educational options depends on one's career motivations. A bachelor's or master's degree may be more appropriate for those seeking to expose themselves to science, but who are not willing to spend a half-decade or more intensely focused scientific research.

Enrolling in a Ph.D., master's, or MBA program with the intention of getting a degree but without being passionate about the topic of study is of relatively little value. The preferred route is to follow your passions, hone your craft, and find ways to apply your expertise to problems you find interesting. Skills learned in school will influence early job roles and career opportunities. Later opportunities will be based upon early performance, and skills and experience learned in earlier positions—being passionate about your job greatly facilitates this growth.

Chapter 20
Final Words

It is absolutely essential to recognize that success comes at
the end of failure after failure after failure … If it were easy, 500
people would have already done it.
Alejandro Zaffaroni

The biotechnology industry was born in 1973 with the development of gene splicing techniques, enabling the directed modification and use of biological systems. The potential to improve drug development methods was quickly recognized as many companies were formed to leverage gene splicing and other related techniques and discoveries. Biotechnology has altered paradigms for drug development, improving efficiencies and enabling new possibilities, and will likewise revolutionize other industries. While many revolutionary products have emerged, companies continue to aggressively research and develop new medical products and technologies.

New discoveries and technologies continue to expand and extend the applications and appeal of biotechnology beyond the prototypical development of new drugs. Biotechnology has improved crop yields, drastically reduced the need for pesticides, and created nutritionally-enhanced foods. Biotechnology also shows great promise for industrial processes: increasing efficiencies of existing processes, reducing waste generation, and enabling unprecedented new possibilities.

This book has presented a history and overview of the biotechnology industry, explaining the scientific, legal, regulatory, political, and commercial factors that shape and define the industry. It is important that the reader develop a holistic view of the industry. One does not need to be an expert in all areas, but it is important to at least appreciate the interplay of factors from disparate elements— the need to consider marketing issues in making R&D decisions, the impact of regulations and politics on marketing, etc.

It is also important to recognize that success in biotechnology requires self-directed lifelong learning. The industry is constantly in motion. Unlike professional fields which often have mandated continuing education programs, in biotechnology it is an individual responsibility. The appendices contain a select set of resources to help maintain exposure to new developments and industry trends.

Armed with this understanding of the drivers of the biotechnology industry and an appreciation of the applications of biotechnology, you are equipped to lead, direct, and profit from the expanding influence of biotechnology.

VI

Appendices

A - Internet Resources

B - Annotated Bibliography

C - Glossary

Appendix A

Internet Resources

NEWS AND INFORMATION

BiotechBlog
http://www.BiotechBlog.com
This blog, managed by *Building Biotechnology* author Yali Friedman, covers new commercial, legal, political, and scientific trends in biotechnology.

Biotechnology@Nature.com
http://www.nature.com/biotech/
The Nature Publishing Group publishes several journals that cover developments and issues in biotechnology. This portal page provides a quick overview of all relevant Nature Publishing Group resources in the field of biotechnology.

Drug Wonks
http://www.drugwonks.com
Drug Wonks is the forum for the Center for Medicine in the Public Interest, covering policy affecting biotechnology. This blog also provides valuable perspectives on many important topics by tracking and responding to Op-Eds and other news items

In the Pipeline
http://pipeline.corante.com/
This blog is written by an active pharmaceutical researcher. His unique insights on the back-stories behind industry developments are an excellent resource to develop a better understanding of the business of

biotechnology.

Law.com
http://www.law.com
This portal for legal professionals includes many articles on biotechnology issues. Type "biotech" or "biotechnology" in the search box for quick access to articles on biotechnology.

Mars Blog
http://blog.marsdd.com/
This blog is hosted by an incubator in downtown Toronto and covers topics such as emerging science and technology, entrepreneurship and business, and innovation policy. This blog is exemplary in its depth of coverage and the demonstrated desire to ask, and address, fundamental questions.

Patent Baristas
http://www.patentbaristas.com/
This blog is a must-visit for interpretation of new regulations, patent rulings, or other leading case developments.

Pharma Marketing Blog
http://pharmamkting.blogspot.com/
This might better be described as a source of how *not* to market. The content frequently addresses marketing gaffes and controversies. An excellent source of guidance for breaking stories and case studies.

Sciencecareers.org Career Development
http://sciencecareers.sciencemag.org/career_development
A career development magazine that helps early-career scientists explore their career options. Research and non-research careers in academia, industry, and elsewhere are explored and profiled.

United States Regulatory Agencies Unified Biotechnology Website
http://usbiotechreg.nbii.gov/index.asp
This website focuses on the agricultural products of modern biotechnology. A searchable database covers genetically engineered crop plants intended for food or feed that have completed all recommended or required reviews for food, feed, or planting use in the United States.

USDA Agricultural Biotechnology Briefing Room
http://www.ers.usda.gov/Briefing/Biotechnology
This Economic Research Service production provides background and coverage of issues on adoption and economics of biotechnology in farming. Topics include marketing, labeling, and segregation issues associated with genetically modified foods and agricultural biotechnology research and development of biotechnology.

U.S. Department of State: Biotechnology
http://usinfo.state.gov/ei/economic_issues/biotechnology.html
Produced by the Office of International Information Programs, this site presents speeches, articles, and links on biotechnology policy, regulations, and science. An excellent resource for international and political issues in biotechnology.

INVESTING AND COMPETITIVE INTELLIGENCE

Biospace
http://www.biospace.com
This resource for general and company-specific biotechnology industry news excels in its investing section. Prominent features include financial figures, company information, regional profiles, a career center, and breaking news.

Drug Patent Watch
http://www.DrugPatentWatch.com
Information on pharmaceutical drug patent expirations, sales statistics, generic equivalents, patent claims, pharmaceutical sponsors, and more. Yali Friedman is founder of Drug Patent Watch.

Recombinant Capital
http://www.recap.com
Advice and analysis related to the environment for corporate and product development and alliance formation. In addition to several value-added databases, Recombinant Capital also produces Signals Magazine (www.signalsmag.com), featuring biotechnology industry analysis.

ANGEL FUNDING

Angel Capital Association Angel Organizations
http://www.angelcapitalassociation.org/dir_directory/directory.aspx

Gaebler Ventures Directory of Angel Investors
http://www.gaebler.com/angel-investor-networks.htm

Inc.com Angel Investor Directory
http://www.inc.com/articles/2001/09/23461.html

VentureChoice Angels Directory
http://www.venturechoice.com/articles/angels_directory.htm

FEDERAL FUNDING

Grants.gov
http://www.grants.gov

SBIR.gov
http://www.sbir.gov/

DEPARTMENT OF AGRICULTURE (USDA)
http://www.csrees.usda.gov/funding/sbir/sbir.html

DEPARTMENT OF COMMERCE
National Oceanic and Atmospheric Administration (NOAA)
http://www.ago.noaa.gov/grants/
National Institute of Standards and Technology (NIST)
http://patapsco.nist.gov/ts_sbir/
NIST Technology Innovation Program
http://www.nist.gov/tip/

DEPARTMENT OF DEFENSE (DoD)
http://www.acq.osd.mil/sadbu/sbir
http://www.dodsbir.net/
Air Force
http://www.sbirsttrmall.com/
Army Research Office
http://www.armysbir.com/

Defense Advanced Research Projects Agency (DARPA)
http://www.darpa.mil/sbir/
Defense Technical Information Center (DTIC)
http://www.dtic.mil/dtic/sbir/
Missile Defense Agency (formerly BMDO)
http://www.winmda.com/
Navy
http://www.navysbir.com/
Special Operations Acquisition and Logistics Center (SOCOM)
http://soal.socom.mil/index.cfm?page=sadbu

DEPARTMENT OF EDUCATION
http://www.ed.gov/programs/sbir/index.html

DEPARTMENT OF ENERGY (DOE)
http://sbir.er.doe.gov/sbir/

DEPARTMENT OF HOMELAND SECURITY
http://www.sbir.dhs.gov/

DEPARTMENT OF TRANSPORTATION
http://www.volpe.dot.gov/sbir/

ENVIRONMENTAL PROTECTION AGENCY (EPA)
http://es.epa.gov/ncer/sbir

DEPARTMENT OF HEALTH AND HUMAN SERVICES
National Institutes of Health, Food and Drug Administration, and Centers for Disease Control
http://grants.nih.gov/grants/funding/sbir.htm

NATIONAL AERONAUTICS & SPACE ADMINISTRATION (NASA)
http://sbir.gsfc.nasa.gov/SBIR/SBIR.html

INDUSTRY ORGANIZATIONS

Biotechnology Industry Organization
http://www.bio.org

Pharmaceutical Researchers and Manufacturers of America
http://www.phrma.org

Appendix B
Annotated Bibliography

SCIENCE

The Billion Dollar Molecule
Barry Werth
Touchstone Books, 1995. ISBN: 0671510576
This book presents a first-hand account of the development of Vertex Pharmaceuticals. Read about the challenges of selecting projects for drug development and the complex interaction of science and business in biotechnology business development.

Biotechnology Journal
Wiley Interscience
http://www.wiley-vch.de/publish/en/journals/alphabeticIndex/2446/
This peer-reviewed journal publishes papers covering novel aspects and methods in all areas of biotechnology, especially those focusing on healthcare, nutrition and technology. Special attention is also paid to the public, legal, ethical and cultural aspects of biotechnological research.

Chemical and Engineering News
American Chemical Society
http://pubs.acs.org/cen/
A weekly magazine, *Chemical and Engineering News* covers many scientific topics in biotechnology. Print subscriptions are free with Society membership and online access is available for non-members.

Genes IX
Benjamin Lewin
Prentice Hall, 2007. ISBN: 0763740632
The textbook of modern molecular biology, *Genes IX* presents current knowledge on the mechanisms of biological processes. The level of discussion is quite advanced; unfamiliar readers may want to complement this with an undergraduate textbook.

Genetic Engineering News
Mary Ann Liebert, Inc.
http://www.genengnews.com
Published 21 times a year, this tabloid-format publication covers the entire bioproduct life cycle, from early-stage R&D, to applied research and bioprocess, through to commercialization, including marketing and regulations. The application scope includes biopharmaceuticals, bio-agriculture, chemicals and enzymes, environmental markets, and emerging biosciences including biodefense, bioenergy, and nanobiotechnology.

Invisible Frontiers
Stephen Hall, James Watson
Oxford University Press, 2002. ISBN: 0195151593
This book tells the story of the race to clone the first human gene, an achievement that led to the formation of Genentech and the birth of biotechnology. Excellent reading for anyone interested in the history and early development of the biotechnology industry.

Modern Drug Discovery
American Chemical Society
http://pubs.acs.org/journals/mdd/
Modern Drug Discovery focuses on emerging trends in drug discovery. A print subscription is free for individuals employed within the drug discovery and/or life science research fields who live in North America, the United Kingdom, or Western Europe.

Nature Biotechnology
Nature Publishing Group
http://www.nature.com/nbt
Nature Biotechnology, a sister publication of the preeminent scientific journal *Nature*, publishes significant application in the pharmaceutical, medical, agricultural, and environmental sciences. Complementing

this function, the journal also features analysis of, and commentary on, published research and business, regulatory, and societal activities that influence this research. A regular supplement series, Bioentrepreneur, provides practical advice on the challenges in building a biotechnology company.

INTELLECTUAL PROPERTY AND REGULATION

From Test tube to Patient
Food and Drug Administration. Fourth Edition, January 2006
http://www.fda.gov/fdac/special/testtubetopatient/
One of the FDA's most popular publications, this report tells the story of new drug development in the United States and highlights the consumer protection role of the Center for Drug Evaluation and Research. Articles describe individual procedures in drug development, from laboratory drug testing to clinical trials and post-marketing surveillance.

Guide to U.S. Regulation of Agricultural Biotechnology Products
Pew Initiative on Food and Biotechnology, 2001
http://pewagbiotech.org/resources/issuebriefs/1-regguide.pdf
This report, focusing on agricultural biotechnology, provides a general overview of the U.S. regulations and laws under which biotechnology products are reviewed for health, safety, and environmental impacts.

IP Management in Health and Agricultural Innovation: A Handbook of Best Practices
A. Krattiger, R.T. Mahoney, L. Nelsen, *et al.*
MIHR: Oxford, UK, and PIPRAL Davis, USA
http://www.iphandbook.org/
This rich guide features 153 chapters by more than 200 authors on practical issues in IP management. Website guides are distill key points in unique contexts designed for policymakers, senior administrators, technology transfer managers, and scientists. A companion blog also provides current commentary on IP management issues.

BUSINESS

The Art of the Start
Guy Kawasaki
Portfolio Hardcover, 2004. ISBN: 1591840562
Written by prolific venture capitalist Guy Kawasaki, *The Art of the Start* provides an overview of the important steps in starting a new venture, including important guidance in refining business plans and pitching investors.

First Fruit
Belinda Martineau
McGraw-Hill, 2001. ISBN: 0071360565
This book profiles the development of the Flavr Savr tomato, the first genetically engineered whole food ever brought to market. Read about the scientific challenges of producing value-added genetically modified plants and the process of satisfying regulatory concerns. Interestingly, Flavr Savr tomatoes did not fail in the market due to public resistance, but rather due to management's inexperience in the premium tomato business which left them unable to sell the tomatoes at a profit.

The Golden Helix
Arthur Kornberg
University Science Books, 1996. ISBN: 0935702326
Nobel laureate Arthur Kornberg was originally skeptical of commercial biotechnology, insisting that academic labs were better equipped to research fundamental issues in biology. In this book, Kornberg details his involvement in the development of Alza, a drug delivery firm, and the growth of the biotechnology industry during this time. Countering his initial sentiments, Kornberg concludes that industry, not academia, is where the most productive science takes place.

Journal of Commercial Biotechnology
Palgrave Macmillan
http://www.palgrave-journals.com/jcb
The *Journal of Commercial Biotechnology* aims to deliver a practical understanding of the strategic development and management associated with the commercialization of biotechnology through dissemination and evaluation of the current techniques, strategic thinking, and best practice in all aspects of the subject. Yali Friedman is managing editor of

the *Journal of Commercial Biotechnology.*

The Journal of Life Sciences
Burrill & Company and the California Healthcare Institute
http://www.tjols.com
The Journal of Life Sciences is a bi-monthly magazine focusing on "where science and society meet," and presents analysis and commentary about the impact of biotechnology and other biosciences on business, policy, and culture.

Science Lessons: What the Business of Biotech Taught Me About Management
Gordon Binder, Philip Bashe
Harvard Business School Press, 2008. ISBN: 9781591398615
This book offers a rare glimpse into the early development of Amgen, one of the biotechnology industry's leading companies. Gordon Binder served in the increasingly central roles of CFO, CEO, and Chairman of Amgen from 1982 to 2000—a span which covers the launch and growth of Epogen. In addition to tracing the development of Amgen and the biotechnology industry, Binder and Bashe also offer practical management advice and recommendations on tackling common business challenges.

Term Sheets & Valuations
Alex Wilmerding
Aspatore Books, 2006. ISBN: 1587620685
An in-depth look at term sheets. In addition to a section-by-section view of a term sheet, valuations, and guidance, this book includes a sample term sheet with a description of each clause and a discussion of key negotiation points and red flags.

U.S. and Canadian Biotechnology VC Directory
BioAbility, LLC
http://www.bioworld.com/
A comprehensive listing of VC firms investing in biotechnology.

Valuation in the Life Sciences: A Practical Guide
Boris Bogdan and Ralph Villiger
Springer, 2007. ISBN: 9783540455653
This comprehensive source for biotechnology valuation offers a mix of

theory and practical examples. A diverse set of valuation examples is covered, helping ground the theory and enabling readers to understand when, and how, to apply valuation methods.

The Use of Biotechnology in U.S. Industries
U.S. Department of Commerce, 2003
http://www.technology.gov/reports/Biotechnology/CD120a_0310.pdf
The first ever in-depth federal government assessment of the development and adoption of biotechnology in industry. This assessment was directed at increasing national policy makers' understanding of the current development and use of biotechnology in U.S. industries, and to assist federal statistical agencies in developing measures and statistics of biotechnology related economic activity.

Glossary

SCIENCE

Absorption, Distribution, Metabolism, Excretion and Toxicology (ADMET): An element of pre-clinical and clinical trials used to measure the effects of a drug on animal and human physiology.

Amino acid: Building block of proteins. Proteins consist of amino acids linked end-to-end. There are 20 different amino acid molecules that make up proteins. The DNA sequence that codes for a gene dictates the order of amino acids in a given protein.

Antibiotic: A chemical substance that can kill or inhibit the growth of a microorganism.

Antibody: Immune system protein produced by humans and higher animals to recognize and neutralize bacteria, viruses, cancerous cells, and other foreign compounds.

Antisense: A natural or synthetic DNA or RNA molecule that specifically binds with messenger RNA to selectively inhibit expression of a single gene.

Applied research: Aimed at gaining knowledge or understanding to determine the means by which a specific recognized need may be met. Applied research builds upon the discoveries of basic research to enable commercialization.

Bacillus thuringiensis: A naturally occurring bacteria that produces Bt toxin, a protein that is toxic to certain kinds of insects. The Bt toxin gene has been genetically engineered into corn and cotton plants to reduce the need for chemical pesticides.

Bacteriophage: Naturally-occuring type of virus that only infect bacteria.

Base: A key component of DNA and RNA molecules. Four different bases are found in DNA: adenine (A), cytosine (C), guanine (G) and thymine (T). In RNA, uracil (U) substitutes for thymine.

Basic research: Aimed at gaining more comprehensive knowledge or understanding of the subject under study, without specific applications in mind. Basic research advances scientific knowledge but does not have specific immediate commercial objectives, although it may be in fields of present or potential commercial interest.

Biofuel: Fuels such as ethanol and diesel produced from sugars, vegetable oils, or other organic matter using biotechnology methods.

Bioinformatics: The application of information technology to manage and analyze the vast amounts of data generated from biological research.

Bioleaching: The use of plants to extract heavy metals from soils.

Bioremediation: The use of biological systems, usually microorganisms, to decompose or sequester toxic and unwanted substances in the environment.

Biotechnology: The application of molecular biology for useful purposes.

Blue biotechnology: A seldom-used term referring to marine and aquatic applications of biotechnology.

Chromosome: The DNA-protein complexes that contain all the genes in a cell.

Cloning: The process of making an identical copy of something. Often used in reference to copying animals, it may also refer to creating copies of DNA fragments, individual cells, or plants.

Codon: A sequence of three DNA or RNA bases that specifies an amino acid in the synthesis of a protein.

Combinatorial chemistry: A product discovery technique that uses robotics and parallel chemical reactions to generate and screen as many as several million molecules with similar structures in order to find chemical molecules with desired properties.

Cytochrome p450: A set of enzymes involved in chemical modification and degradation of chemicals including drugs and other foreign compounds.

Data mining: Using computers to analyze masses of information to discover trends and patterns.

Diagnostic: A product used for the diagnosis of a disease or medical condition.

DNA (deoxyribonucleic acid): The primary source of genetic information in cells. DNA is comprised of nucleotides and is composed of two strands wound around each other, called the double helix.

DNA fingerprinting: A DNA analysis method that measures genetic variation among individuals. This technology is often used as a forensic tool to detect differences or similarities in blood and tissue samples at crime scenes.

DNA sequencing: The process of determining the exact order of bases in a segment of DNA.

Double-blind: An experimental protocol whereby neither the experimental subjects nor the administrators know whether a drug or placebo is being administered. Double-blind protocols are used to eliminate bias.

Drug delivery: The process by which a formulated drug is administered to the patient.

Drug development: The process of taking a lead compound, demonstrating it to be safe and effective for use in humans, and preparing it for commercial-scale manufacture.

Enzyme: A functional protein that catalyzes (speeds up) a chemical reaction. Enzymes control the rate of naturally occurring metabolic processes such as those necessary for growth and reproduction.

***Escherichia coli* (*E. coli*):** A common gut bacteria that is a workhorse and model organism for molecular biology.

Excipient: An inactive ingredient (there are no absolutely inert excipients) added to a drug to give it a pill form or otherwise aid in delivery.

Expression: A highly specific process in which a gene is switched on at a certain time and its encoded protein is synthesized, resulting in the manifestation of a characteristic that is specified by a gene. Genetic predispositions to disease arise when a person carries the gene for a disease but it is not expressed.

False negative: An experimental outcome that incorrectly yields a negative result. False negatives can complicate disease diagnosis.

False positive: An experimental outcome that incorrectly yields a positive result. False positives can frustrate assessing the performance of lead compounds.

Fermentation: Technically the process of breaking complex organic substances into simpler ones, such as conversion of sugars into alcohols, acetone, or lactic acid. Also refers to any large-scale cultivation of microbes or other single cells (e.g., for drug production).

Functional genomics: The use of biological experiments and genetic correlations to establish what each gene does, how it is regulated, and how it interacts with other genes.

Functional foods: Foods containing compounds with beneficial health effects beyond those provided by the basic nutrients, minerals, and vitamins.

Gene: The fundamental unit of heredity, a segment of DNA which encodes a defined biochemical function. Some genes direct the synthesis of proteins, while others have regulatory functions.

Gene expression: The production of a gene product—generally defined as the synthesis of an encoded protein.

Gene splicing: Splicing a gene from one segment of DNA into another. Commonly used to insert foreign genes into bacteria for analysis, or to insert foreign genes into bacteria or other organisms for genetic modification or to produce and harvest large quantities of specific proteins.

Gene therapy: The replacement of a defective gene in a person or organism suffering from a genetic disease.

Genetic code: The language in which DNA's instructions are written. The genetic code consists of triplets of nucleotides (codons), with each triplet corresponding to one amino acid in a protein structure, or a signal to start or stop protein production.

Genetic disorder: A condition or mutation that results from one or more defective genes.

Genetic engineering: The manipulation of genes to create heritable changes in biological organisms and products that are useful to people, living things, or the environment.

Genetic predisposition: A susceptibility to disease that is related to a genetic condition, which may or may not result in actual development of the disease.

Genetic screening: The use of a specific biological test to screen for inherited diseases or medical conditions.

Genome: The sum of an organism's genes.

Genomics: The study of genes and their function.

Good manufacturing practice (GMP): Guidelines ensuring the quality and purity of chemical products that are intended for use in pharmaceutical applications, and controls ensuring that methods and facilities used for production, processing, packaging, and storage result in drugs with consistent and sufficient quality, purity, and activity.

Gray biotechnology: A seldom used term for industrial applications of biotechnology. More commonly referred to as white biotechnology.

Green biotechnology: The use of biotechnology for agricultural applications.

Human Genome Project: The international research effort which identified and located the full sequence of bases in the human genome.

Incidence: measure of the rate of new occurrences of a disease or condition in a population.

Immune system: The cells, biological substances (such as antibodies), and cellular activities that work together to recognize foreign substances and provide resistance to disease.

In silico (in computer): Computer-based predictions that can complement *in vitro* and *in vivo* procedures.

In vitro (in glass): Experimental procedures carried out in test-tubes, beakers, etc.

In vivo (in the living body): Experimental procedures carried out on living cell lines or in living animals.

Lead compound: In pre-clinical development and clinical trials, a potential drug being tested for safety and efficacy.

Liposome: An artificial membrane. Can be used to encapsulate drugs and aid in drug delivery.

Microarray: A tool that permits the identification of DNA samples and examination of gene expression in individual tissues and different conditions.

Monoclonal antibody: A synthetic immune system protein that recognizes a single target. Polyclonal antibodies recognize several related targets.

Molecular evolution: The process of making discrete changes in genes to improve the functional characteristics of proteins and enzymes.

Molecular farming: Using biotechnology to produce useful products from domesticated plants and animals.

mRNA (messenger RNA): A ribonucleic acid molecule that transmits genetic information from DNA to the protein synthesis machinery in cells, where it directs protein synthesis.

Mutant: A cell or organism harboring one or more mutated genes.

Mutation: A change in the base sequence of a gene that results in it not performing its normal task.

Nanotechnology: A technology field focusing on materials at sizes measured in billionths of a meter.

Nucleotide: One of the structural components, or building blocks, of DNA and RNA. A nucleotide consists of a base plus one molecule of sugar and phosphoric acid.

Oncogenic: Viruses, chemicals, genes, proteins, etc. that cause the formation of tumors.

Pathogen: A disease-causing organism.

Personalized medicine: The practice of medicine in which therapies are developed for and directed at the patients most likely to benefit from them.

Pharmacogenetics: Examination of the differences in drug response between individuals—one drug, many genomes.

Pharmacogenomics: Examination of differences in how one person responds to different drugs—many drugs, one genome.

Pharming: The process of farming genetically engineered animals and plants to produce drugs.

Placebo: A mock-treatment used in single-blind or double-blind experiments to eliminate bias from experiment subjects or administrators, respectively.

Platform technology: A technique or tool that enables a range of scientific investigations. Examples include combinatorial chemistry for producing novel compounds, microarrays for gene expression analysis, and bioinformatics programs for data assembly and analysis.

Polymerase Chain Reaction (PCR): A method to produce sufficient DNA for analysis from a very small amount of DNA.

Prevalence: measure of how commonly a disease or condition occurs in a population.

Prion: A naturally occurring protein that can be converted into a disease-causing form. Prion diseases can be transmitted in the absence of DNA or RNA.

Promoter: A DNA sequence preceding a gene that contains regulatory sequences influencing the expression of the gene.

Proof-of-principle: Demonstration of the commercial potential of a discovery or invention.

Protein: A long-chain molecule comprised of amino acids that folds into a complex three-dimensional structure. The type and order of the amino acids in a protein is specified by the nucleotide sequence of the gene that codes for the protein. The structure of a protein determines its function.

Proteomics: The study of the protein profile of each cell type, protein differences between healthy and diseased states, and the function of, and interaction among, proteins.

Rational drug design: Using the known three-dimensional structure of a molecule, usually a protein, to design a drug that will bind have a therapeutic effect on it.

Recombinant DNA: The DNA formed by combining segments of DNA from different sources.

Red biotechnology: The use of biotechnology for therapeutic applications.

Reformulation: Altering an established drug's formulation or delivery method to yield improvements in safety or efficacy.

Repurposing: Finding new indications for approved drugs.

Restriction enzyme: A protein that cuts DNA molecules at specific sites, dictated by the nucleotide sequence.

Retrovirus: A type of virus that reproduces by converting RNA into DNA.

Single Nucleotide Polymorphism (SNP): A single base difference in the sequence of a gene which alters the structure and function of the gene product.

RNA (ribonucleic acid): A nucleic acid, similar to DNA, which has roles in gene expression.

RNA interference: Using antisense techniques to selectively inhibit expression of a gene.

Stem cell: An undifferentiated cell that can multiply and become any sort of cell in the body.

Telomere: The tip of a chromosome. Telomeres are involved in the replication and stability of chromosomes.

Tissue engineering: The production of natural or synthetic organs and tissues that can be implanted as fully functional units or may develop to perform necessary functions following implantation.

Transcription: The synthesis of an mRNA molecule as a copy of a gene. In gene expression, transcription precedes translation.

Translation: The synthesis of a protein based on the nucleotide sequence of an mRNA molecule, which corresponds to the sequence of a gene.

Transgenic: An organism with one or more genes that have been transferred to it from another organism.

Vaccine: A preparation of either whole disease-causing organisms (killed or weakened) or parts of such organisms, used to confer immunity against the disease that the organisms cause. Vaccine preparations can be natural, synthetic, or derived by recombinant DNA technology.

White biotechnology: The use of biotechnology for industrial applications.

X-ray crystallography: An essential technique for determining the three-dimensional structure of biological molecules.

Xenotransplantation: Transplanting a foreign tissue into another species.

LEGAL

Claim: A comprehensive and precise description that defines the scope of an invention.

Compulsory license: A license in which a government forces the holder of a patent or other exclusive right to grant use to the state or others. Authorized under World Trade Organization provisions to enable countries to produce generic versions of patented drugs in the event of a health crisis.

Continuation: A filing, while a patent is active, which contains additions or changes to the previous claims.

Copyright: The exclusive legal right to publish, perform, display, or distribute an original work.

Divisional patent: A patent that covers the same specification as a previous (parent) patent, but claims a different invention.

Ex parte: A legal proceeding where only one party is represented. Patent prosecution is an *ex parte* procedure.

Experimental use: The practice of a patented invention solely with intention of experimentation or perfection of the invention.

Evergreening: The practice of launching new formulations, combinations, delivery methods, and indications for drugs facing patent expiration to effectively increase the duration of patent-protected sales.

Filing date: The date on which a complete patent application is received by the Patent and Trademark Office.

Freedom to operate: The absence of intellectual property and regulatory impediments (which may require patent license and passage of enabling laws) to commercialization.

***Inter partes* reexamination:** A method by which third parties challenge the validity of a patent on the grounds of prior art publication without resorting to litigation.

Interference: When two or more patent applications or issued patents claim the same invention.

License: An agreement whereby one party gains access to another's technology (e.g. a patent license).

Non-disclosure agreement (NDA): An agreement, common between companies and their contractors and partners, which allows a company to share protected information while preventing its release.

Office action: A formal response by a patent examiner regarding a patent application or amendment.

Patent: A description of an invention. Patents contain one or more claims that describe the subject matter covered in sufficient detail to permit skilled experts to practice an invention, and grant the right to exclude others from practicing an invention.

Patent agent: An individual with technical training who is capable of representing an inventor in patent prosecution.

Patent attorney: An individual with legal training in patent law who is capable of representing an inventor in patent prosecution and litigation.

Patent pool: An agreement between two or more patent owners to license one or more of their patents to one another or third parties.

Patent term adjustment: Provisions to adjust patent term to provide restoration for U.S. patent and trademark office delays.

Patent term extension: Provisions to extend patent term to account for time spent waiting for FDA approval.

Prior art: Public knowledge that exists in a field; all previously issued patents, publications, public announcements, or knowledge that bear on the invention claimed in a patent application.

Prosecution: The process by which an inventor engages with the patent office to obtain a patent and determine the scope of its claims.

Provisional patent application: A preliminary patent application filed without a formal patent claim, oath or declaration, or any prior art statement. It provides the means to establish an early effective filing date in a subsequent non-provisional patent application and allows the term "Patent Pending" to be applied.

Reach-through claim: A patent claim to rights to royalties from, or rights to use, drugs or other physical or intellectual property produced using the patent

Submarine patent: A patent that emerges after it has unknowingly been infringed upon.

Trade secret: Knowledge and information that is not generally known to the industry. Examples include customer lists, business plans, and manufacturing methods.

Trademark: A registered name, word, symbol, or device identifying a company's products or services.

REGULATORY

Abbreviated New Drug Application (ANDA): A simplified submission permitted for a generic version of an approved drug.

Accelerated approval: A process to make products for life threatening diseases available on the market prior to formal demonstration of benefit. Uses surrogate markers—indirect measures of efficacy—and requires continued testing to confirm efficacy.

Action letter: An official FDA communication that informs the sponsor of an NDA or BLA of a decision by the agency. An approval letter allows commercial marketing of the product.

Authorized generic: Drugs produced by branded companies and marketed under a private label to compete with other generic drugs.

Bayh-Dole Act: Provides the statutory basis and framework for federal technology transfer activities, including patenting and licensing federally funded inventions to commercial ventures.

Bioequivalence: Demonstration that a generic drug has the same chemical and biological properties as its pioneer counterpart.

Biologic: Medicine made by biological processes rather than by chemical synthesis or extraction. Biologics typify biotechnology-derived drugs. Contrast with small-molecule drugs.

Biologics License Application (BLA): Application filed with the FDA Center for Biologics Evaluation and Research (CDER) for approval to market a biologic drug.

Biosimilar: A generic biologic drug that is "similar but not identical" to a pioneer drug.

Brand-name drug: The original, often patented, version of a drug. Contrast with generic drugs.

Clinical pharmacology study: Clinical trial designed to determine the absorption, distribution, metabolism, elimination, and toxicity (ADMET) of a drug.

Clinical trial: A human study designed to measure the safety and efficacy of a new drug.

Current good manufacturing practices (cGMP): Regulatory practices to ensure safety and consistency of manufacturing processes.

Exclusivity: A temporary FDA-granted monopoly, distinct from patent or other intellectual property protection. Exclusivity may be granted for developing drugs for rare diseases, novel drugs, conducting pediatric clinical trials, or successfully challenging invalid patents.

Fast track: A process for interacting with the FDA during drug development, intended for drugs to treat serious or life threatening conditions that demonstrate the potential to address an unmet medical need.

First-in-man study: Phase I trial primarily concerned with establishing the safety of a compound.

Follow-on biologic: An FDA term for a biologic drug that is similar to an existing biologic.

Generic drug: The version of an approved drug produced by a competitor after a pioneer firm's patents expire.

Hatch-Waxman safe harbor: A research-use exemption stemming from the Hatch-Waxman Act which exempts from infringement the use of patented inventions in preparation for submitting drug applications.

Hatch-Waxman Act: Contains provisions to foster the development of generic drugs and support pioneer drug development.

Indication: A use for which a specific drug is approved by the FDA.

Institutional Review Board (IRB): An independent committee of scientists, physicians, and lay people that oversees clinical trials.

Investigational New Drug (IND): An application to pursue clinical trials with an experimental drug that has passed pre-clinical development.

March-in rights: A stipulation of the Bayh-Dole Act enabling the government to request and potentially require issuance of a license to a patent, which was developed with federal funding, to another party.

Named Patient Program: European compassionate use program, enabling limited distribution of drugs prior to approval.

New Drug Application (NDA): Application filed with the FDA Center for Drug Evaluation and Research (CDER) for approval to market a small-molecule drug.

Off-label use: Use of a drug not in accordance with FDA-approved uses or drug labeling. Physicians are free to prescribe drugs for off-label uses.

Orange Book: Also known as *Approved Drug Products with Therapeutic Equivalence Evaluations*, the *Orange Book* contains detailed information on all approved drugs and must list all extant patents.

Orphan Drug: A drug that treats a disease affecting fewer than 200,000 Americans or for which there is no reasonable expectation that the cost of research and development will be recovered from sales in the United States. The Orphan Drug Act provides special incentives for producers of orphan drugs.

OTC-switch: The process of gaining approval to sell a drug over the counter, which may grant 3 years exclusivity for the over-the-counter market.

Over the counter (OTC): Selling a drug without a prescription. Requires evidence that patients can self-diagnose and use the drug safely without physician supervision.

Phase I: Clinical trial designed primarily to determine the safety of an experimental drug.

Phase II: Clinical trial that evaluates an experimental drug's safety, assesses side effects, and establishes dosage guidelines.

Phase III: Clinical trial designed to assess the safety and effectiveness of an experimental drug. Success in Phase III trials can lead to marketing approval.

Phase IV: Post-approval clinical trials used to monitor safety and efficacy or examine additional applications of drugs.

Pioneer (brand-name) drug: The patented version of a drug. Contrast with generic drugs, the competing versions produced when pioneer patents expire.

Pre-clinical studies: Studies that test a drug on animals and nonhuman test systems. Safety information from such studies is used to support an investigational new drug application (IND).

Reverse payment: A payment from a branded drug company to a generic drug company to delay launch of a generic drug.

Salami slicing: Filing for multiple orphan drug designations on the same drug.

Small-molecule drug: A drug produced using defined chemical synthesis or extraction. Contrast with biologics, drugs produced by biological processes.

Surrogate marker: An indirect measure of effectiveness, such as a laboratory test or tumor shrinkage, used to show a strong potential for effectiveness in accelerated drug approval.

COMMERCIAL

Accredited investor: A type of investor, largely defined by their wealth, permitted to invest in high-risk investments.

Acquisition: Appropriation of the controlling interests of one company by another.

Alliance: Agreement between two or more companies to cooperate in some way.

Angel investor: Wealthy individual who personally provides startup capital to very young companies to help them grow.

Barrier to entry: A condition that makes it difficult for competitors to enter the market (e.g., patent, trademark, high up-front capital requirements).

Blockbuster: Drug with $1 billion or more in sales.

Board of directors: A group legally charged with the responsibility to protect the interests of a company and its shareholders.

Bootstrap: Starting a business with little or no external funding.

Bridge loan: A short-term, high-interest, loan provided to companies in dire need of cash.

Burn rate: The rate at which an unprofitable company is going through its available cash reserves.

Business model: A description of a company's purpose, commercial offerings, strategies, organizational structure, operational processes, etc. Often confused with business plan, below.

Business plan: A formal statement of a company's goals and the plan for reaching those goals.

Comparable: A valuation technique based on analogy to similar companies or products.

Competitive advantage: An advantage that a firm has relative to competing firms; may be in the form of intellectual property, expertise, partnerships, assets, etc.

Controlling interest: Ownership of more than 50 percent of a company's voting shares.

Convertible: Securities (usually bonds or preferred shares) that can be converted into common stock.

Cooperative research and development agreement (CRADA): An agreement enabling federally funded laboratories to perform for-profit contract work for commercial firms.

Corporate inversion: Formation of a parent corporation of a U.S. company in a country with little or no corporation tax, and structuring a U.S. subsidiary to manage U.S. sales. This scheme enables foreign sales to be taxed in local markets only, and not be taxed in the U.S.

Cross-licensing: An agreement in which two or more firms with competing and similar technologies strike a deal to reduce the need for legal actions to clarify who is to profit from applications of the technology.

Dilution: The decrease in relative ownership among existing investors as additional shares are issued.

Discounted cash flow (DCF): A valuation technique that attempts to consider future events and determine a present value for a product or project based on the variety of outcomes.

Discovery rights: Selling only research findings while retaining rights to knowledge discovered in the course of research and development.

Down round: A financing event in which a company is valued lower than it was previously.

Due diligence: The process by which research is conducted to determine the value of an investment, licensing agreement, merger, or other similar activity.

Earnings strippings: A potential result from a corporate inversion where payments from a U.S. subsidiary to a foreign parent are used as deductions against U.S. taxes.

Elevator pitch: A short summary—typically less than two-minutes—used to quickly describe a business to investors

Dumb money: Funding from investors who cannot provide additional benefits such as guidance or networking.

Equity dilution: The dilution of the equity stakes of founders and early investors by subsequent investments.

Equity investment: An investment purchasing partial ownership of a company.

Exit: The means by which investors gain a return on their investment, commonly through sale of shares in public markets or acquisition by another company.

Free cash flow: The amount of cash available to a company after all expenses have been paid.

Friends and family: A term for investments which often help start a company, and are typically made by unaccredited investors.

Incubator: A facility offering space and shared services and facilities to early-stage companies.

Initial Public Offering (IPO): The initial sale of shares of a private company in public markets, turning it into a publicly-traded company.

Institutional support : Esteem granted to companies by their affiliation with highly regarded partners, financiers, and other affiliates.

Intellectual property: Intangible assets such as patents, trade secrets, trade names, etc.

License: An agreement to grant rights to a patent or tangible subject.

Market segmentation: The division of a market into distinct groups of buyers or decision makers.

Medicaid: Government-subsidized healthcare coverage for individuals with low incomes and limited resources.

Medical tourism: Travelling for medical treatment. Often motivated by cost-savings, local waiting lists and expertise availability, or local regulatory restrictions.

Medicare: Government-subsidized healthcare for individuals 65 years of age and older, some disabled people under 65 years of age, and people with permanent kidney failure.

Merger: The formal combination of two companies into one entity. Often used to refer to acquisitions. A merger can be distinguished from consolidation, in which a new separate entity is created.

Mezzanine funding: Funding that generally leads to liquidity (IPO or merger) or commercial launch and eventual profitability.

Milestone: The completion of a specified phase in product development. Investors and alliance partners may use milestones to establish a timeline for incremental investments or payments.

Offshoring: The relocation of business processes from one country to another.

Options-pricing: A valuation technique that analyzes the value of discrete operational paths.

Outsourcing: The execution and management of selected operations by outside parties.

Parallel trade: The trade of products between countries without permission of the intellectual property owner. Often used to capitalize on inter-country price differences.

Pharmacoeconomics: Study of the cost-benefit ratios of drugs.

PIPE (Private Investment in Public Equity): Purchase of discounted shares in a public company in which payment goes directly to the company rather than to existing shareholders.

Preferred stock: A convertible offering that cannot be sold until it is converted to common stock.

Price elasticity: A measure of the change in demand resulting from a change in price of a product or service. Low price elasticity indicates little change in demand; high elasticity indicates a relatively large change in demand.

Price/earnings (P/E) ratio: A rudimentary technique to determine the value of a company, or relative value of several companies, by comparing the share price to annual earnings.

Private equity: In contrast to owning shares in a public company, private equity is ownership in a private company.

Proxy fight: A process by which shareholders can vote for corporate changes. May be used to appoint new directors or replace senior management.

Royalty: The payment of a percentage of sales as compensation to product developers, patent licensors, or even investors.

Ratchet: An anti-dilution provision where an investor is granted additional shares of stock without charge if the company later sells the shares at a lower price.

Return on Investment (ROI): Profit (or loss) on an investment, often expressed as a percentage.

Reverse merger: The merger of a private company with a "shell" company, rendering the private company public.

SBIR (Small Business Innovation Research): A funding program that encourages small business to explore their technological potential and provides the incentive to profit from its commercialization.

Scientific Advisory Board (SAB): A group of esteemed scientists and business professionals, independent from management, which provides objective feedback and guidance on a company's progress and goals.

Seed financing: Capital furnished to prove the feasibility of a concept or invention.

Secondary offering: A public or private share offering subsequent to an initial public offering.

Series A/B/C/D: Venture funding stages that fund product development and early commercial launch activities.

Shell: A public company with few or no assets that may be the remnant of a bankruptcy or asset sale. Used in reverse mergers to enable a private company to become public.

Smart money: Funding from investors who are able to contribute guidance, networking, or other benefits.

Special purpose acquisition company (SPAC): A public company formed with the intent of engaging in a reverse merger with a private company.

Spin-off: Separating a smaller unit from an established company, permitting each company to retain focus while shielding the parent from risk and granting the spin-off the administrative benefits of small size.

STTR (Small Business Technology Transfer): A funding program that encourages public/private sector partnership in order to develop new technologies and profit from their commercialization.

Targeted marketing: The alignment of marketing efforts with the benefits sought by individual market segments.

Tax arbitrage: Location of specific operations in countries and regions with favorable tax treatment for those operations.

Technology transfer: The transfer of discoveries made by basic research institutions, such as universities and government laboratories, to the commercial sector for development into useful products and services.

Venture capital: Money invested by venture capitalists in startup companies in exchange for equity.

Venture capitalist: An individual who invests in start-up companies with the intent of making a large return on investment.

Virtual company: Firms that outsource all or most of the elements of research, development, and marketing.

Index

About the Author

Yali Friedman is managing editor of the *Journal of Commercial Biotechnology* and serves on the science advisory board of Chakra Biotech and the editorial advisory boards of the *Biotechnology Journal, Journal of Medical Marketing* and *Open Biotechnology Journal.* He regularly guest-lectures for biotechnology education programs, teaching classes on the business of biotechnology, and has written and given talks on diverse topics such as biotechnology entrepreneurship, strategies to cope with a lack of management talent and capital when developing companies outside of established hubs, and new paradigms in technology-based economic development.

Yali also has a long history in biotechnology media, having created a *Forbes* "Best of the Web"-rated web site on the biotechnology industry for a NY Times company and managed it for many years. His other projects include the Student Guide to DNA Based Computers, sponsored by FUJI Television, BiotechBlog.com, and DrugPatentWatch.com, a pharmaceutical industry competitive intelligence service.

Yali can be contacted at *info@thinkbiotech.com.*

Related titles from Logos Press
www.logos-press.com/books

BEST PRACTICES IN BIOTECHNOLOGY BUSINESS DEVELOPMENT
Eleven chapters from biotechnology industry experts
ISBN: 978-09734676-0-4

BEST PRACTICES IN BIOTECHNOLOGY EDUCATION
22 chapters on programs from 5 countries
ISBN: 978-09734676-7-3

Printed in the United States
204131BV00003B/1-63/P

9 780973 467666